AI-Powered Agriculture

The Next Green Revolution

By
Drew Ashton

AI-Powered Agriculture

The Next Green Revolution

Table of Contents

Introduction 1

Chapter 1: The Dawn of AI in Agriculture 5
    The Evolution of Farming Technologies............................... 5
    The Role of AI in Modern Agriculture 9

Chapter 2: Understanding AI and Machine Learning 13
    Basics of AI and Machine Learning 13
    How AI Technologies Apply to Farming............................ 17

Chapter 3: Precision Agriculture... 21
    Definition and Benefits 21
    Real-time Data Collection and Analysis............................ 25

Chapter 4: Crop Monitoring and Management.................... 29
    AI in Crop Surveillance 29
    Disease Detection and Prevention 33

Chapter 5: Soil Analysis and Health..................................... 37
    AI-driven Soil Testing.. 37
    Optimizing Soil Health for Better Yields 41

Chapter 6: Water Management 45
    AI Solutions for Efficient Irrigation 45
    Monitoring Water Usage and Soil Moisture 49

Chapter 7: Weed and Pest Control 53
    Identifying and Managing Weeds with AI 53
    AI for Pest Prediction and Management 57

Chapter 8: Autonomous Machinery 61
   AI-powered Tractors and Harvesters 61
   Robots in Planting and Maintenance 65

Chapter 9: Predictive Maintenance 69
   AI for Equipment Maintenance and Repair 69
   Reducing Downtime with Predictive Analytics 73

Chapter 10: Supply Chain Optimization 77
   Enhancing Logistics with AI 77
   Improving Storage and Distribution 81

Chapter 11: Market Analysis and Decision Making 85
   AI in Market Predictions 85
   Data-driven Decision Making 89

Chapter 12: Smart Greenhouses 93
   Integrating AI in Controlled Environments 93
   Optimizing Growth Conditions 97

Chapter 13: Livestock Management 102
   AI in Tracking Animal Health and Productivity 102
   Automated Feeding and Milking Systems 106

Chapter 14: Sustainability and Environmental Impact 110
   Reducing Carbon Footprint with AI 110
   Promoting Biodiversity and Conservation 114

Chapter 15: Economic Benefits of AI in Agriculture 118
   Cost Reduction and Efficiency Gains 118
   Economic Challenges and Considerations 121

Chapter 16: Policy and Regulation 124
   Navigating AI-related Agricultural Policies 124
   Ethical Considerations and Data Privacy 128

Chapter 17: Global Case Studies 132
   Successful Implementations Around the World 132

Lessons Learned from Diverse Climates and Cultures 136

Chapter 18: AI in Small-Scale Farming 141
Tools and Technologies for Small Farmers 141
Empowering Smallholder Farmers 145

Chapter 19: Challenges and Limitations 149
Technological and Practical Hurdles 149
Addressing Skepticism and Resistance 152

Chapter 20: Future Trends in AI Agriculture 156
Emerging Technologies and Innovations 156
Preparing for the Future of Farming 160

Chapter 21: Educating the Next Generation 164
AI Curriculum for Agriculture 164
Training Programs and Workshops 168

Chapter 22: Collaborations and Partnerships 172
Role of Governments and NGOs 172
Industry-Academia Partnerships 175

Chapter 23: Funding and Investment 180
Securing Funding for AI Projects 180
Investment Opportunities in AgriTech 183

Chapter 24: AI and Climate Change 188
Mitigating Climate Impact through AI 188
Adapting to Changing Environmental Conditions 193

Chapter 25: Community Engagement and Social Impact 197
Engaging Local Communities 197
Social Benefits of AI in Agriculture 201

Conclusion 206

Appendix A: Appendix 209
Data Sets and References 209

Formulas and Equations 210
Software Tools and Applications 210
Additional Reading and Resources 210
Contact Information 211

Glossary of Terms 212

Introduction

Welcome to a new era in agriculture, one that merges the time-honored practice of farming with cutting-edge technology. As we stand on the brink of the next green revolution, it's clear that artificial intelligence (AI) is not just a futuristic concept; it's here, and it's transforming the way we grow food. From smallholder farms in developing countries to expansive corporate fields, AI is reshaping agriculture in ways that were once unimaginable.

Imagine a world where crops are monitored by drones, soil health is analyzed in real-time, and irrigation systems are optimized for maximum efficiency. This isn't science fiction; it's modern farming, powered by AI. The potential for AI to enhance agricultural practices is immense. It promises to increase yields, reduce waste, and make farming more sustainable—all of which are critical as the global population continues to grow.

At its core, this book seeks to educate and inspire. Whether you're a tech enthusiast fascinated by AI, a farmer eager to improve your practices, or someone simply curious about how food makes its way to your table, there's something in these pages for you. We will traverse the expansive landscape of AI in agriculture, breaking down complex topics into understandable segments and showcasing real-world applications that illustrate the technology's tangible benefits.

AI is not a monolithic technology; it's a collection of diverse tools and techniques. Machine learning, computer vision, robotics, and data analytics all fall under the AI umbrella, each contributing uniquely to agricultural innovation. For instance, in precision agriculture, AI algorithms analyze data from various sources—such as satellite imagery and local weather stations—to recommend optimal planting and harvesting times. In crop monitoring, AI-powered drones survey fields and identify signs of stress or disease that might go unnoticed by the human eye.

But let's take a step back. It's important to understand the journey that has brought us to this point. The agricultural sector has always been a hotbed of innovation, from the advent of the plow to the mechanization of the industrial age. And now, AI represents the next leap forward, a leap that will demand a new level of adaptability and learning from all stakeholders.

One can't discuss AI in agriculture without addressing the challenges and limitations. As promising as this technology is, its integration into farming practices is not without hurdles. Issues such as data privacy, the cost of implementation, and the need for new skill sets among farmers are significant. However, these challenges also present opportunities for growth and innovation. Solutions, collaborations, and new business models are emerging to make AI accessible and beneficial to all.

For those working in agriculture, knowledge is power. The more we understand about the technology and its applications, the better positioned we will be to harness its potential. This book is structured to provide that foundational knowledge and beyond. We'll explore the basics of AI and machine learning, delve into precision agriculture, and examine the specifics of crop monitoring, soil health, water management, and more.

The global agricultural community stands to gain immensely from AI. Farmers in different climates and regions face unique challenges, and the adaptability of AI solutions means that practices can be tailored to meet local needs. From enhancing crop resilience in arid regions to reducing pesticide use in more temperate zones, the benefits are manifold. Real-life case studies will highlight successful implementations around the world, offering lessons and insights that can be applied in diverse settings.

Moreover, AI has the potential to address some of the most pressing environmental issues of our time. Agriculture is a major contributor to greenhouse gas emissions, but AI can help mitigate this impact. Through smarter resource management and optimized supply chains, farming can become more sustainable and eco-friendly. The implications for biodiversity, soil health, and water conservation are profound, making AI an essential tool in the fight against climate change.

One of the most exciting aspects of AI in agriculture is its inclusivity. Small-scale and large-scale farms alike can benefit. Advanced technologies are becoming more affordable and easier to use, leveling the playing field for farmers of all sizes. This democratization of technology is crucial for global food security, enabling smallholder farmers to improve their yields and livelihoods.

As we look to the future, it's clear that AI will continue to evolve and integrate more deeply into agricultural systems. Emerging technologies like blockchain, Internet of Things (IoT), and advanced robotics will further enhance the capabilities of AI. The agricultural workforce will need to adapt, embracing new skills and knowledge areas to stay relevant in this rapidly changing landscape.

Education and training will play a pivotal role in this transition. By fostering a culture of continuous learning and innovation, we can prepare the next generation of farmers and tech developers to make the most of AI. Workshops, training programs, and academic curriculums focused on AI in agriculture will be essential in nurturing this talent pool.

Partnerships and collaborations are also crucial. Governments, NGOs, private enterprises, and academic institutions all have a role to play. By working together, they can create ecosystems that support the development and deployment of AI technologies in agriculture. Funding and investment will be key drivers, enabling the scaling of innovations and ensuring that benefits are widespread and impactful.

This book aims to be a comprehensive guide for anyone interested in the transformative power of AI in agriculture. It's about fostering a deeper understanding of the technology, its applications, and its potential. And more importantly, it's about inspiring action. The future of farming is bright, and AI is a significant part of that future. By embracing this change, we can forge a path toward more sustainable, efficient, and productive agricultural practices.

So, embark on this journey with an open mind and a forward-looking spirit. The chapters that follow will lead you through the intricacies of AI in agriculture, revealing the innovations that are reshaping the fields, farms, and food systems worldwide. Let's explore the incredible intersection of artificial intelligence and agriculture and discover together what the future holds for this vital industry.

Chapter 1:
The Dawn of AI in Agriculture

The first rays of the artificial intelligence revolution in agriculture are just beginning to illuminate what will surely be a transformative era. For centuries, farming has been defined by manual labor and traditional knowledge, but the rise of AI is set to redefine these paradigms. AI's role in modern agriculture is not merely an augmentation of existing practices but a complete overhaul that promises unprecedented efficiency, sustainability, and productivity. With AI, the age-old challenges of weather unpredictability, pest management, and crop yield optimization can be tackled with data-driven precision. This new dawn heralds a future where farmers are empowered with real-time insights and predictive analytics, paving the way for a more resilient and productive agricultural sector. The synergy between cutting-edge algorithms and time-honored farming techniques is poised to drive the next green revolution, much needed in our ever-growing world.

The Evolution of Farming Technologies

Farming has a rich and varied history, evolving from simple hand tools to complex machines across millennia. At the heart of this progression has always been humankind's relentless drive to increase productivity, efficiency, and sustainability in food production. From the rudimentary ploughs of ancient times to the sophisticated, GPS-guided tractors of today, each wave of technological advancement has transformed the agricultural landscape.

In the earliest days of agriculture, tools were rudimentary but revolutionary. The advent of simple implements like hoes and sickles marked the beginning of a more structured approach to cultivation. These tools allowed early farmers to sow and reap with greater efficiency than ever before, which, in turn, led to the establishment of more stable food sources and the growth of civilizations.

The next significant leap came with the domestication of animals for farming purposes. Oxen and horses became essential parts of agricultural work, substantially increasing the area that could be farmed and the speed at which it could be done. The use of animal labor represented a seismic shift in farming capabilities, allowing for larger-scale agricultural practices and contributing to the growth of expansive agricultural societies.

As history moved forward, the Industrial Revolution heralded another monumental shift with the introduction of mechanized farming equipment. Steam engines powered the first mechanical reapers and threshers, making it possible to harvest crops on a previously unimaginable scale. These machines reduced dependency on manual labor and animal power, which, in turn, slashed costs and elevated productivity.

The 20th century brought yet another transformative era characterized by the development of tractors and combine harvesters. These innovations integrated multiple functions into single, versatile machines, making farming operations more streamlined and efficient. With the combination of motor engines and hydraulic devices, tractors became the backbone of modern farming, making it feasible to cultivate and harvest massive areas of land with unprecedented speed and precision.

Advances in chemical science during this time also revolutionized agriculture. Synthetic fertilizers, pesticides, and herbicides became ubiquitous, boosting yields and crop resilience. However, these innovations came with their own sets of challenges, including environmental impact and the potential for long-term soil degradation. This led to the search for more sustainable farming practices, which has been ongoing into the present day.

In recent decades, digital technology has made its mark on agriculture through the introduction of Geographic Information Systems (GIS) and remote sensing. These technologies have allowed for precise mapping and monitoring of large agricultural areas, enabling farmers to optimize planting patterns, irrigation schedules, and pest control measures. The ability to gather and analyze vast amounts of data has opened new avenues for increasing efficiency and conserving resources, laying the groundwork for the integration of AI in agriculture.

All these evolutionary steps have set the stage for the current revolution driven by artificial intelligence. AI represents a culmination of centuries of technological advancements, offering unprecedented levels of precision, efficiency, and sustainability. By leveraging machine learning algorithms, AI can analyze vast datasets far beyond human capability, identifying patterns and making predictions that can optimize every aspect of farming from planting to harvesting.

One of the most exciting developments in AI-driven agriculture is precision farming, which employs AI technologies to analyze factors such as soil conditions, weather patterns, and crop health in real-time. This level of granular insight allows for highly customized farming strategies that can significantly boost yields while minimizing resource input. Sensors and IoT devices provide continuous data streams that feed into AI systems, making it possible to make instant, informed decisions.

Moreover, AI is revolutionizing labor-intensive tasks through automation. Autonomous tractors and harvesters equipped with AI can operate independently, further increasing efficiency and reducing the need for human labor. These machines are not only faster and more accurate than their human counterparts, but they also operate tirelessly, ensuring that agricultural activities can continue around the clock.

Crop monitoring and disease detection are other areas where AI is having a significant impact. Using machine learning algorithms, AI systems can analyze images captured by drones or satellites to identify signs of disease, nutrient deficiencies, or pest infestations long before they become visible to the naked eye. This preventative approach can save entire crops, improving food security and reducing waste.

As we look toward the future, the integration of AI in agriculture is poised to address some of the most pressing challenges facing our planet. With climate change threatening traditional farming practices, AI offers adaptive solutions that can help farmers mitigate the effects of extreme weather and shifting environmental conditions. By optimizing resource usage and reducing waste, AI-driven technologies support sustainable farming practices that are essential for feeding a growing global population.

Water management is another critical area where AI technologies are making a difference. Precision irrigation systems driven by AI ensure that crops receive the exact amount of water they need, thereby conserving this precious resource. Advanced sensors can monitor soil moisture levels and weather forecasts, enabling intelligent irrigation schedules that adapt to real-time conditions.

The journey from hand tools to AI-driven machinery is a testament to human ingenuity and adaptability. Each evolutionary step in farming technology has built upon the innovations that preceded it, driving agriculture toward a future where it can meet the demands of an ever-growing global population. AI is not just a continuation of this journey; it represents a transformational leap that holds the promise of a new green revolution.

The story of farming technologies is one of relentless progress and adaptation, with each new development opening up new possibilities for productivity and sustainability. As AI continues to evolve, its integration into agriculture will likely lead to even more groundbreaking advancements, transforming farming into a highly efficient, data-driven industry that can meet the challenges of the 21st century and beyond.

The evolution of farming technologies is far from over. As we move forward, continuous innovation will be essential to address the emerging challenges and maximize the potential of AI in agriculture. By embracing these technological advancements, we can create a more sustainable, productive, and resilient agricultural system that holds the promise of feeding future generations while preserving our planet.

The Role of AI in Modern Agriculture

Artificial Intelligence (AI) has fundamentally transformed various sectors, profoundly impacting agriculture. This integration is more than just a technological upgrade; it's a paradigm shift that propels farming into a new era of efficiency, sustainability, and productivity. Gone are the days when decisions were made based solely on intuition or past practices. Today, AI gives farmers the power of precise, data-driven insights, making farming smarter and more resilient.

One of the cornerstone aspects of AI in modern agriculture is its capability to enhance decision-making processes. By leveraging machine learning algorithms, vast amounts of data collected from various sources such as sensors, satellites, and drones can be analyzed in real-time. This empowers farmers to predict crop yields, identify potential threats, and optimize resource allocation with remarkable accuracy. Imagine having the ability to foresee the impact of weather patterns or pest invasions well in advance – AI makes this possible.

Crop monitoring is another critical area where AI proves invaluable. Traditional methods of crop surveillance often involved manual inspections, which were time-consuming and prone to human error. Nowadays, AI-powered drones and imaging technologies can scan large fields in minutes, providing detailed insights into plant health, growth stages, and areas requiring attention. This transition from manual to automated monitoring enables farmers to cover more ground with less effort, ensuring that every aspect of the crop's lifecycle is meticulously tracked.

The benefits of AI don't stop at crop monitoring. Advanced analytics can process data on soil health, water usage, and nutrient levels. AI systems can recommend the right amount of fertilizer needed for each section of the field, minimizing wastage and environmental impact. For instance, AI algorithms analyze soil samples to determine nutrient deficiencies and then cross-reference this with other data sets such as weather forecasts and crop requirements. This enables precise nutrient management strategies tailored to specific conditions, thereby optimizing soil health and boosting yield.

A key advantage brought by AI is the efficiency in resource utilization. Efficient water management is crucial in agriculture, given the global challenges of water scarcity and climate change. AI-driven irrigation systems use real-time data to optimize water distribution across fields. This ensures that crops receive the exact amount of water they need, reducing wastage and conserving a critical resource. Drones equipped with thermal cameras can also identify variations in soil moisture levels, providing invaluable data that helps in making informed irrigation decisions.

Another arena where AI makes waves is in pest and weed control. Traditional methods of pest management often involved blanket pesticide applications, leading to significant ecological impact and resistance buildup in pests. AI systems identify specific pest threats through image recognition and suggest targeted interventions. This ensures that only the affected areas receive treatment, thereby promoting sustainable pest management practices. Similarly, AI-powered systems can differentiate between crops and weeds, allowing for precise weed control solutions that minimize the use of herbicides.

Beyond the fields, AI also revolutionizes post-harvest processes. From sorting and grading crops to predicting market trends, artificial intelligence streamlines operations across the agricultural supply chain. Machine learning models analyze market data, consumer behaviors, and historical trends to forecast market demands. This enables farmers to make informed decisions on crop planting, harvesting, and distribution, ultimately leading to reduced wastage and higher profitability.

Furthermore, AI plays a pivotal role in enhancing the longevity and efficiency of agricultural machinery. Predictive maintenance systems utilize AI to monitor equipment performance in real-time, identifying potential issues before they result in costly breakdowns. Sensors embedded in machinery collect data on operational variables such as temperature, vibration, and usage patterns. AI algorithms analyze this data to predict wear and tear, scheduling maintenance activities proactively. This not only extends the lifespan of machinery but also ensures that farming operations run smoothly without unexpected downtime.

The benefits of AI in agriculture extend to environmental sustainability. Sustainable farming practices, enabled by AI, reduce the carbon footprint of agricultural activities. Precision agriculture techniques minimize the over-application of fertilizers and chemicals, limiting their runoff into water bodies and reducing greenhouse gas emissions. AI-driven monitoring systems help conserve biodiversity by maintaining balanced ecosystems in agricultural landscapes. For instance, AI-powered conservation tools can assess the impact of farming practices on local wildlife and suggest measures to mitigate negative consequences, thereby promoting ecological harmony.

Despite the numerous advantages, the adoption of AI in agriculture is not without challenges. Access to technology, data privacy concerns, and the need for training are significant hurdles that need to be addressed. However, the potential of AI to drive the next green revolution cannot be understated. Empowering farmers with the tools and knowledge to integrate AI into their operations is essential for maximizing its benefits. Governments, educational institutions, and private organizations must collaborate to provide the necessary infrastructure and support, ensuring that AI-driven agriculture is accessible to all, from large commercial farms to smallholder farmers in remote regions.

The transformative impact of AI in modern agriculture offers a glimpse into the future of farming. With continuous advancements in technology and an increasing emphasis on sustainability, AI will play an even more crucial role in shaping the agricultural landscapes of tomorrow. The journey towards a smarter, more efficient, and sustainable farming ecosystem has only just begun, and the integration of AI is a beacon of hope illuminating the path forward.

Chapter 2:
Understanding AI and
Machine Learning

In the journey to harness artificial intelligence (AI) for transformative agricultural advancements, it's essential to grasp the fundamental concepts of AI and machine learning (ML). AI refers to the simulation of human intelligence processes by machines, particularly computer systems, whereas ML is a subset of AI focusing on the development of algorithms that permit computers to learn from and make predictions based on data. Both AI and ML operate in synergy, offering remarkable capabilities to analyze vast datasets, recognize patterns, and optimize farming practices based on predictive insights. Imagine a system that can forecast weather changes, recommend the best planting strategies, and even identify early signs of crop diseases—all through sophisticated algorithms continuously improving from real-time data. Understanding these technologies is the first step to unlocking their potential, paving the way for smarter, more sustainable farming practices that can address global food security challenges and catalyze the next green revolution.

Basics of AI and Machine Learning

The fields of Artificial Intelligence (AI) and Machine Learning (ML) have been gaining traction in various sectors, and agriculture is no exception. At their core, AI and ML involve using algorithms and computational methods to analyze data, identify patterns, and make decisions without human intervention. This foundation serves as the stepping stone to understanding how these technologies can revolutionize modern farming.

AI encompasses a broad array of techniques aimed at imitating human intelligence, ranging from simple rule-based systems to more complex machine learning models. Meanwhile, ML is a subset of AI that empowers systems to learn from data and improve over time without explicit programming. In essence, where AI strives to mimic human cognition, ML focuses on learning and self-improvement.

Grasping the basics of AI and ML begins with understanding data. Data is the fuel that drives these technologies. Whether it's satellite imagery, soil moisture levels, weather conditions, or crop health indicators, the vast amounts of data generated on farms can be harnessed through AI and ML to make more informed decisions. Efficient data collection and management form the crux of any AI/ML system and are vital for producing accurate and actionable insights.

Machine Learning operates through models—computational algorithms that learn from historical data. These models can be classified into different types like supervised learning, unsupervised learning, and reinforcement learning. In supervised learning, the model is trained on a labeled dataset, which means the input-output pairings are known, enabling the system to predict outcomes based on new data. For instance, predicting crop yield based on historical weather patterns and soil information.

Unsupervised learning, on the other hand, deals with unlabeled data. The goal here is to find hidden patterns or intrinsic structures in the input data. For example, identifying clusters of crops that are more likely to have similar watering needs based on soil composition and weather variations. Reinforcement learning involves a system interacting with its environment to attain certain goals, refining its strategy through feedback from its actions. This technique is useful in autonomous machinery where tractors and drones learn to navigate fields more efficiently over time.

Artificial Neural Networks (ANNs) are one of the most effective machine learning techniques employed today. Inspired by the human brain, ANNs consist of layers of interconnected nodes, or neurons, that process data in hierarchical stages. When applied to agriculture, ANNs can predict crop diseases, optimize irrigation schedules, and even forecast market demands. Such tasks involve processing large volumes of data, recognizing complex patterns, and making decisions that can significantly impact productivity and sustainability.

Another crucial aspect is the concept of training and testing datasets. The dataset is divided into two parts: one is used to train the model, and the other to test it. This approach helps verify the model's accuracy and generalizability. A well-trained model can make reliable predictions based on new, unseen data, thus enabling farmers to take proactive measures.

Support Vector Machines (SVMs), Decision Trees, and Random Forests are other machine learning methods that hold significance in agriculture applications. SVMs excel in classification tasks like identifying plant species or disease infection stages from images. Decision Trees and Random Forests are useful for tasks that require understanding variations, such as predicting the influence of different agricultural practices on crop yield. These techniques provide the decision-making framework AI needs to identify the most impactful variables and actions.

Machine learning models are only as good as the data they are trained on. Therefore, data quality, relevance, and diversity are crucial. Collecting data across different seasons, crop types, and geographic regions can help create robust models that generalize well across various conditions. This requires collaboration between farmers, agronomists, and data scientists to implement effective data-gathering protocols and share knowledge.

Natural Language Processing (NLP), another AI subset, also has potential in agriculture. NLP can analyze textual data, such as research papers, weather forecasts, market reports, and even social media, to glean insights that might be valuable for farmers. For instance, understanding the sentiment around market trends or integrating diverse information sources into a cohesive advisory system can help farmers make better-informed decisions.

Computer Vision is yet another AI technology making waves in agriculture. By automating visual identification tasks, computer vision can monitor crop health, detect pests, and estimate yields. Coupling this with drone and satellite imagery enables large-scale, real-time farm monitoring. The combination of high-resolution imagery and advanced algorithms ensures precision in identifying areas that need attention, thus allowing targeted interventions.

AI and ML also provide the groundwork for predictive analytics, enabling foresight into future events. Predictive models can forecast weather patterns, crop yields, and even potential pest outbreaks, giving farmers the information they need to plan effectively. This foresight is invaluable for mitigating risks, optimizing resource allocation, and maximizing yields.

Another innovation driven by AI and ML is the concept of digital twins. Digital twins are virtual replicas of physical entities—in this case, farms. By simulating various scenarios, digital twins can offer insights into the potential outcomes of different farming practices, enabling farmers to experiment and optimize their strategies in a risk-free environment.

It's worth noting that the implementation of AI and ML in agriculture isn't without its challenges. Data privacy, ethical concerns, and the need for significant initial investments are hurdles that need addressing. Additionally, the complexity of creating and deploying AI solutions means that interdisciplinary collaboration is essential. Agronomists need to work closely with data scientists and technologists to ensure the solutions are practical and scientifically sound.

In sum, understanding the basics of AI and Machine Learning involves appreciating the value of data, the variety of algorithms available, and the potential for these technologies to adapt and improve. These foundational elements empower AI and ML to transform agriculture into a more efficient, sustainable, and profitable endeavor. By embracing these technologies, farmers can significantly enhance productivity while reducing environmental impact, setting the stage for the next green revolution.

How AI Technologies Apply to Farming

Artificial Intelligence (AI) technologies are not just revolutionizing the tech industry—they're changing the face of agriculture too. At the core, AI in farming means utilizing machine learning algorithms and other advanced data analytics to enhance efficiency, accuracy, and sustainability. From weather forecasting to automating mundane tasks, AI-powered systems offer solutions to many of the age-old challenges in farming.

One of the most critical applications of AI in farming is in data collection and analysis. Machine learning models can analyze vast amounts of data from various sources such as satellite imagery, weather data, and soil sensors. This data is crucial for making informed decisions, whether it's planning the planting schedule or optimizing irrigation. Farmers traditionally relied on experience and observation; AI augments this, offering data-backed insights that help increase yields and minimize waste.

AI-driven predictive analytics also plays a significant role. By analyzing historical data along with current conditions, AI models can forecast potential challenges such as pest infestations or disease outbreaks. These predictive capabilities allow farmers to take preemptive measures. For instance, if there's a high probability of a pest outbreak, farmers can treat their crops with pesticides in advance, thereby reducing the damage and saving resources.

In addition to data analysis, AI technologies are transforming hardware used in agriculture. Autonomous tractors and harvesters can now navigate fields with pinpoint accuracy, driven by GPS technology and AI algorithms. These machines reduce the need for manual labor and can operate around the clock, significantly enhancing productivity. Additionally, robot-assisted planting systems ensure seeds are planted at optimal depths and spacing, which is critical for germination and growth.

Precision agriculture is an area where AI demonstrates its profound impact. By leveraging AI algorithms, farmers can deploy resources such as water, fertilizers, and pesticides more efficiently. AI can recommend the exact quantities needed in specific areas, reducing waste and lowering costs. This tailored approach ensures that crops receive the nutrients they need when they need them, boosting overall productivity and sustainability.

Moreover, AI technologies are instrumental in monitoring crop health. Traditional methods of crop surveillance are labor-intensive and often lack precision. AI-based solutions like drones equipped with sensors and cameras can monitor vast fields quickly and accurately. These drones capture high-resolution images that are analyzed by AI models to identify signs of disease, nutrient deficiencies, or pest infestations. By catching these issues early, farmers can mitigate damage and improve crop health.

In the realm of soil management, AI contributes to better understanding and stewardship of this vital resource. AI-powered soil sensors can provide real-time data on soil moisture, nutrient levels, and pH. Machine learning models analyze this data to offer insights and recommendations for improving soil health. For example, AI can suggest crop rotations or soil amendments to maintain soil fertility, leading to better yields over time.

Water management is another critical area where AI technologies are making a difference. Traditional irrigation systems often over-water or under-water crops, leading to wasted resources and reduced crop yields. AI-driven irrigation systems use real-time data to optimize watering schedules and amounts. These systems take into account various factors like weather forecasts, soil moisture levels, and crop needs to ensure efficient water use.

It's also worth noting the role of AI in livestock management. AI technologies enable farmers to monitor animals' health and behavior more closely than ever before. Wearable sensors and cameras collect data on vital signs and activity levels. AI models analyze this data to detect anomalies that could indicate health issues, ensuring that livestock receive timely care. This not only improves animal welfare but also boosts productivity and profitability.

Supply chain optimization is another frontier where AI proves invaluable. From forecasting demand to managing inventory levels, AI helps streamline processes, reducing waste and ensuring that produce reaches consumers in the best possible condition. AI can analyze supply chain data to identify inefficiencies and recommend improvements, leading to quicker, more reliable delivery of fresh produce.

In the context of market analysis and decision-making, AI technologies provide actionable insights that help farmers navigate market fluctuations. AI models can predict market trends, allowing farmers to make informed decisions about what crops to plant and when to sell them. This helps in maximizing profits and reducing the risk of financial losses due to market volatility.

AI isn't just about large-scale farming; it's equally transformative for smallholder farmers. Affordable, AI-powered tools are being developed to support small-scale farming operations. These tools can help farmers make better decisions, improve crop yields, and enhance their livelihoods. By providing access to cutting-edge technology, AI levels the playing field, offering opportunities for smaller farms to thrive.

The integration of AI in smart greenhouses is another exciting development. These controlled environments benefit immensely from AI's ability to monitor and manage growth conditions. AI systems regulate variables such as temperature, humidity, and light to create optimal growth conditions. This leads to higher yield rates and reduced resource consumption compared to traditional farming methods.

Sustainability and environmental impact are at the heart of AI's contributions to farming. By optimizing resource use, reducing waste, and improving efficiency, AI technologies help make agriculture more sustainable. They contribute to reducing the carbon footprint of farming activities and promote practices that conserve biodiversity and protect natural ecosystems.

In conclusion, the application of AI technologies in farming is multifaceted and transformative. From predictive analytics and autonomous machinery to precision agriculture and livestock management, AI offers innovative solutions that enhance efficiency, productivity, and sustainability. As these technologies continue to evolve, they hold the promise of driving the next green revolution, transforming agriculture for a more food-secure and environmentally friendly future.

Chapter 3:
Precision Agriculture

Precision agriculture represents a revolutionary approach to farming, leveraging advanced technologies and data analytics to optimize crop management and resource use. By collecting real-time data from various sources, such as soil sensors, weather stations, and satellite imagery, farmers can make informed decisions that enhance productivity and sustainability. This method enables precise application of water, fertilizers, and pesticides, reducing waste and minimizing environmental impact. Precision agriculture not only boosts yields but also promotes economic efficiency, allowing farmers to achieve more with less. As we dive deeper into precision agriculture, we'll uncover how these innovative practices are transforming the agricultural landscape, paving the way for a more sustainable and technologically advanced future in farming.

Definition and Benefits

Precision agriculture, often referred to as site-specific crop management, represents a paradigm shift in how we grow and manage our crops. At its core, precision agriculture is an advanced farming technique that utilizes data analytics, AI, and various sensors to monitor and optimize agricultural processes with pinpoint accuracy. By leveraging technology, farmers can make informed decisions that lead to improved efficiency, higher yields, and more sustainable farming practices.

One might ask, what exactly falls under the domain of precision agriculture? In essence, it involves utilizing AI-driven tools and methods such as GPS-guided tractors, drone surveillance, soil sensors, and satellite imagery to obtain real-time data about the condition of crops and the surrounding environment. This rich data pool allows farmers to tailor their practices to meet the specific needs of each plot of land, enhancing productivity while minimizing resource waste.

One of the immediate benefits of precision agriculture is the optimization of resource use. Traditional farming methods often rely on uniform application of water, fertilizer, and pesticides across an entire field. However, this approach can lead to overuse in some areas and underuse in others, resulting in inefficiencies and increased costs. Precision agriculture, on the other hand, allows for variable rate application, meaning that inputs are applied exactly where and when they are needed, thereby optimizing the use of resources and reducing costs substantially.

Another significant advantage of precision agriculture is its capability to reduce environmental impact. By applying fertilizers and pesticides only where necessary and in the appropriate amounts, farmers can drastically reduce runoff into local water systems, thereby minimizing pollution and promoting a healthier ecosystem. Additionally, precision irrigation techniques ensure that water is used efficiently, conserving this precious resource and mitigating the effects of droughts and water scarcity.

The ability to monitor crop health in real-time also cannot be overstated. With AI and machine learning algorithms analyzing data from sensors, drones, and satellites, farmers can detect signs of disease, pest infestation, or nutrient deficiencies much earlier than traditional methods would allow. Early detection is critical for prompt intervention, reducing the risk of severe crop loss and ensuring higher quality yields. This aspect of precision agriculture significantly enhances overall farm profitability and sustainability.

Precision agriculture also plays a crucial role in elevating the quality of produce. By closely monitoring and adjusting the growing conditions, farmers can ensure that crops reach their full potential. This not only increases the yield but also improves the nutritional content and market value of the produce. Enhanced crop quality contributes to better health outcomes for consumers and opens up new market opportunities for farmers, particularly those who aim to meet stringent quality standards.

Moreover, the data-driven approach of precision agriculture helps in making better-informed decisions. Farmers can analyze data over time to discern patterns and trends, which can guide future planting and harvesting decisions. For instance, understanding soil composition and moisture levels at a granular level can help farmers decide which crops are best suited for a particular piece of land. This strategic planning contributes to efficient land use and higher overall productivity.

Adopting precision agriculture also means embracing innovation and staying competitive in a quickly evolving marketplace. With the agricultural sector facing increasing pressure to produce more with less, driven partly by a growing global population and changing climate conditions, innovation is not just an advantage but a necessity. Farmers who integrate precision agriculture techniques position themselves at the forefront of the agri-tech revolution, ensuring their long-term viability and success.

Additionally, the scalability of precision agriculture technologies makes them suitable for farms of all sizes. While large-scale commercial farms often lead the way in adopting cutting-edge technologies, small and medium-sized farms can also benefit. The modular nature of many precision agriculture tools allows farmers to start small, perhaps with a single drone or a few soil sensors, and gradually expand their tech arsenal. This scalability ensures that the benefits of precision agriculture are accessible to a broad spectrum of the farming community.

One might also consider the broader social and economic benefits of precision agriculture. By making farming more efficient and sustainable, it supports rural economies and can create new business opportunities. Technology companies that develop precision agriculture tools, for instance, provide jobs and contribute to local and national economies. Additionally, more efficient and profitable farming can help to stabilize food prices and improve food security, which is vital for both developing and developed nations.

However, the societal impacts don't stop at economic and environmental gains. Precision agriculture can also contribute to reducing the labor intensity associated with traditional farming methods. By automating tasks such as planting, monitoring, and harvesting, these technologies can alleviate labor shortages and allow farm workers to focus on more skilled tasks. This transformation can lead to improved working conditions and a higher quality of life for those involved in the farming sector.

Finally, the wealth of data generated by precision agriculture provides a valuable resource for ongoing research and development. Scientists and engineers can use this data to develop even more effective and efficient technologies, creating a feedback loop that continually advances the field of precision agriculture. As these technologies evolve, they will undoubtedly bring about new and unforeseen benefits, further transforming agriculture in ways that can only be anticipated with excitement and optimism.

In conclusion, the definition and benefits of precision agriculture extend far beyond mere technological advancement. It is a holistic approach to farming that incorporates data analytics, AI, and smart technologies to create a more efficient, sustainable, and profitable agricultural system. Precision agriculture represents not just the future of farming but a necessary step toward addressing the global challenges of food security, environmental sustainability, and economic stability in the agricultural sector.

Real-time Data Collection and Analysis

In the landscape of modern agriculture, real-time data collection and analysis stand as pillars supporting the edifice of precision farming. By leveraging advanced technologies, farmers can obtain and interpret data with unprecedented speed and accuracy. It essentially allows for a granular, moment-to-moment understanding of agronomic conditions, enabling more informed and rapid decision-making.

The backbone of real-time data collection is an intricate network of sensors and IoT (Internet of Things) devices strategically embedded in the farming environment. These devices measure a multitude of variables, such as soil moisture, temperature, humidity, light levels, and even the presence of specific nutrients. When placed in a well-planned grid or pattern, they send continuous streams of data to a central system. Software algorithms then analyze this raw data to extract meaningful insights.

Consider, for instance, a typical crop field equipped with soil moisture sensors. These sensors continuously monitor water content across different sections of the field. If an arid patch is detected, automated irrigation systems can be triggered to deliver precise amounts of water to that specific location, thereby conserving water and promoting uniform crop growth. Such precision is impossible with traditional methods, which apply water, fertilizer, and pesticides uniformly across the entire field, often leading to wastage and suboptimal crop yields.

However, the scope of real-time data collection goes beyond soil and water management. Drones equipped with multispectral imaging cameras can capture high-resolution images of the crops. These images are then processed using advanced algorithms to detect plant health, disease outbreaks, and even pest infestations long before they are visible to the naked eye. Immediate actions can be taken, like applying localized treatments, to mitigate these issues, thus minimizing crop loss and maximizing efficiency.

The importance of real-time data in understanding microclimates cannot be overstated. Farms are not monolithic entities; they contain diverse microenvironments that can vary dramatically in terms of temperature, humidity, and other climatic factors. Understanding these variations allows for precise interventions tailored to each microenvironment's specific needs. For instance, a particular section of a vineyard might have a microclimate that fosters optimal grape growth, while another section might be prone to diseases. Targeted interventions can then be planned to ensure uniform grape quality across the vineyard.

The fusion of AI and machine learning with real-time data collection has further amplified its potential. Machine learning models can ingest vast amounts of data and identify complex patterns that humans would struggle to discern. These models can predict future conditions, such as the likelihood of disease outbreaks or optimal harvest times, based on historical and real-time data. This predictive capability allows farmers to plan proactively, thus stabilizing yields and reducing risks.

Additionally, cloud computing has revolutionized how real-time data is stored, processed, and accessed. Farmers no longer need high-end computing hardware on-site; instead, they can store their data in the cloud and utilize cloud-based analytical tools. This democratization of technology makes advanced farming techniques accessible even to small-scale farmers, who can gain valuable insights into their farming operations without hefty upfront investments in technology.

But it's not just about gathering data—it's about translating that data into actionable insights. User-friendly dashboards and mobile applications interpret complex datasets into easy-to-understand metrics and visualizations. Farmers can access these insights on their smartphones or tablets, even while they are out in the field. Such accessibility ensures that decisions can be made swiftly and accurately, embodying the real-time aspect of modern agricultural practices.

Beyond individual farms, real-time data collection has significant implications for regional agricultural planning. Aggregated data from multiple farms can provide a comprehensive understanding of regional agricultural patterns, such as emerging disease trends or shifts in climate conditions. Governments and agricultural bodies can use this aggregated data to design targeted interventions, distribute resources more effectively, and implement regulations based on real-world data.

Collaborative platforms also play a pivotal role in this ecosystem. Shared databases where farmers, researchers, and agronomists contribute and access data can drive community learning and collective problem-solving. For instance, a disease pattern identified in one region can alert other regions to take preventive measures. This collective intelligence network ensures that the benefits of real-time data collection extend beyond individual farms to impact broader agricultural practices globally.

Moreover, integrating blockchain technology with real-time data collection can enhance the traceability and transparency of agricultural produce. From farm to fork, every step can be meticulously documented in an immutable ledger. Consumers can scan a product's QR code to access its entire lifecycle data, from the field conditions during its growth to the transportation methods used. This transparency builds consumer trust and adds value to agricultural products.

The road to widespread implementation of real-time data collection in agriculture is not without its challenges. Issues such as data privacy, high initial setup costs, and the need for technical know-how can pose significant barriers. However, ongoing advancements in technology and a growing emphasis on educational initiatives are steadily overcoming these obstacles, paving the way for broader adoption.

Indeed, real-time data collection and analysis represent a quantum leap forward in the quest for sustainable and efficient agriculture. It's about cultivating not just crops, but knowledge—real-time insights that lead to smarter farming practices, resource conservation, and ultimately, higher yields. As we continue to innovate and refine these technologies, the vision of a highly efficient and eco-friendly agricultural future becomes ever more attainable.

In conclusion, real-time data collection and analysis are revolutionizing the agricultural sector. By providing accurate, timely insights, these technologies empower farmers to make informed decisions that optimize resource use and maximize crop yields. The ongoing fusion of AI, IoT, and cloud computing promises to unlock even greater potentials, heralding a new era of precision farming that could very well be the cornerstone of the next green revolution.

Chapter 4:
Crop Monitoring and Management

Crop monitoring and management, now powered by sophisticated AI technologies, has become a cornerstone of thriving modern agriculture. Through the integration of advanced sensors and data analytics, farmers can keep a vigilant eye on their crops, ensuring optimal growth conditions. AI-driven crop surveillance systems provide real-time insights into plant health, hydration levels, and nutrient needs, transforming raw data into actionable information. With disease detection algorithms, early identification of potential threats to crop health is possible, enabling timely interventions that can prevent widespread damage. This dynamic marriage of AI and agriculture not only promises increased yields and reduced waste but also fosters a more sustainable approach to farming. By leveraging these cutting-edge tools, the agricultural sector stands on the cusp of a new era, equipped to meet the challenges of a growing global population and changing environmental conditions.

AI in Crop Surveillance

In today's rapidly evolving agricultural landscape, crop surveillance plays a pivotal role in ensuring optimal yield and quality. Thanks to the advancements in artificial intelligence, surveillance has taken a giant leap forward. AI-driven crop surveillance is not merely about observing crops; it's about integrating a complex system of technologies to monitor, predict, and manage agricultural practices with unprecedented precision.

AI in crop surveillance entails a suite of technologies that collectively work to monitor various parameters within fields. These include the health status of plants, pest infestations, water levels, and nutrient adequacy though we scope it under general monitoring and management here. One of the central components of AI in crop surveillance is the deployment of sensor networks and imaging technologies. Advanced drones equipped with high-resolution cameras and multispectral sensors fly over fields, capturing detailed images and data. This information is then processed using sophisticated AI algorithms that can distinguish subtle differences in plant health and growth patterns.

Machine learning models play a crucial role in analyzing the data collected from these surveillance systems. By training these models on vast datasets, they become adept at identifying patterns and anomalies that might indicate pest infestations, diseases, or nutrient deficiencies. For instance, a machine learning model can analyze the color and texture of plant leaves to detect early signs of diseases such as blight or mildew long before they become visible to the human eye. This early detection enables farmers to take timely actions, thereby preventing the spread of diseases and minimizing crop loss.

One of the most remarkable aspects of AI-driven crop surveillance is its ability to provide real-time insights. Conventional methods of crop monitoring are often reactive, relying on periodic manual inspections that can miss critical early warning signs. In contrast, AI systems continuously scan and analyze the fields, offering real-time alerts and recommendations. Imagine a farmer receiving a notification on their smartphone that a particular section of their field is showing signs of water stress. Such timely information allows for immediate corrective measures, ensuring that crops receive the right amount of water at the right time.

The integration of AI in crop surveillance also facilitates precision agriculture, a farming management approach that uses data-driven techniques to optimize field-level management regarding crop farming. Precision agriculture aims to ensure that crops and soil receive exactly what they need for optimal health and productivity. AI aids in achieving this by providing detailed maps and models of fields, highlighting areas that might require different treatment. For instance, parts of a field that are nutrient-deficient can be identified and treated with targeted fertilizer applications, thereby reducing waste and cost while improving yield.

Beyond mere observation, AI-driven crop surveillance systems also incorporate predictive analytics. By analyzing historical data and current environmental conditions, these systems can forecast future trends and potential challenges. For example, based on weather forecasts, soil conditions, and crop health data, AI models can predict the likelihood of pest infestations or disease outbreaks. This predictive capability empowers farmers to adopt preventive measures rather than merely reacting to problems as they arise.

Moreover, AI in crop surveillance is pivotal in facilitating sustainable farming practices. By optimizing resource use and minimizing waste, AI systems contribute to the environmental sustainability of agricultural operations. For instance, precision irrigation systems driven by AI ensure that water is used efficiently, reducing wastage and conserving a valuable resource. Similarly, targeted fertilizer applications minimize the risk of runoff and environmental contamination.

Additionally, the accessibility and affordability of AI technologies are expanding rapidly, making them increasingly available to small and medium-sized farms. Traditionally, advanced agricultural technologies were often only within reach for large-scale commercial farms. However, the proliferation of affordable sensors, drones, and cloud-based AI services is democratizing access to these state-of-the-art tools. Smallholder farmers can now leverage AI-driven crop surveillance to boost their productivity and profitability, leveling the playing field in the agricultural sector.

Another intriguing aspect of AI in crop surveillance is its potential to integrate with other technologies like blockchain for enhanced transparency and traceability in supply chains. For instance, data collected from AI surveillance systems can be logged on a blockchain, providing verifiable records of crop conditions, treatments, and yields. This integration not only ensures quality and safety for consumers but also adds value for farmers by creating trust and transparency in their products.

Such sophisticated surveillance mechanisms necessitate an intricate interplay of hardware and software. The hardware encompasses drones, IoT sensors, and imaging devices, while the software includes AI algorithms and machine learning models. Together, they create a robust surveillance system capable of handling enormous amounts of data. The role of cloud computing can't be overstated in this context. Cloud platforms offer the computational power and storage capabilities required to process and analyze the vast amounts of data generated. Furthermore, cloud-based AI applications enable farmers to access and analyze data remotely, breaking down geographical barriers and allowing for real-time decision-making no matter where they are located.

While the benefits of AI in crop surveillance are manifold, the implementation does come with its set of challenges. Data privacy and security are significant concerns, as the vast amounts of data collected can be sensitive. Farmers and agricultural businesses must navigate the complexities of data ownership and privacy regulations to ensure the safe and ethical use of AI technologies. Moreover, there is a learning curve inherent in adopting new technologies. Farmers must be equipped with the necessary skills and knowledge to effectively use AI-driven tools. Educational programs, training initiatives, and extension services play a crucial role in helping farmers transition to AI-enabled agriculture.

In summary, AI in crop surveillance is transforming the way we monitor and manage crops. By providing real-time insights, predictive analytics, and precision farming capabilities, AI is ushering in a new era of agriculture that is more efficient, sustainable, and profitable. The technology's ability to integrate seamlessly with other modern agricultural practices and its potential to democratize access to cutting-edge tools makes it a game-changer for farmers of all scales. As we continue to innovate and improve these systems, the future of farming looks brighter and more resilient than ever.

Disease Detection and Prevention

The health of crops is paramount for achieving optimal yields and ensuring food security. Traditional methods of crop disease detection typically entail manual scouting, which is labor-intensive, time-consuming, and prone to human error. AI, with its unparalleled ability to analyze vast amounts of data rapidly and accurately, is revolutionizing this critical aspect of crop management by introducing robust, automated disease detection and prevention mechanisms.

AI-powered tools are capable of scrutinizing crop fields continuously, capturing real-time data through various sensors and cameras. Machine learning algorithms then process this data to identify disease symptoms at early stages, even before they become visible to the human eye. This early detection is not just a technological marvel but a practical necessity. An early diagnosis can prevent the spread of disease to other parts of the field, saving both crops and costs.

One of the most promising applications of AI in disease detection involves the use of computer vision technology. High-resolution cameras mounted on drones or robotic devices traverse the fields, capturing images of crops from various angles and perspectives. These images are then compared against a vast dataset containing millions of images of healthy and diseased plants. Advanced neural networks sift through these pixels, recognizing patterns and anomalies that indicate the presence of disease.

Deep learning models take this a step further by not only identifying the disease but also classifying its type and suggesting appropriate treatments. For instance, a model may recognize a specific fungal infection and recommend a precise fungicide, thus reducing the blind use of pesticides and promoting more sustainable farming practices.

The integration of AI in disease detection doesn't just stop at visual inspection. In-situ sensors that measure environmental conditions like humidity, temperature, and soil moisture levels contribute valuable data. These physical parameters significantly affect plant health and play a crucial role in disease proliferation. Machine learning algorithms correlate these environmental variables with the likelihood of disease outbreaks, thus offering predictive insights.

Imagine a scenario where an AI system alerts a farmer about a potential disease outbreak two weeks in advance, suggesting preventive measures like adjusting irrigation schedules or applying organic treatments. This kind of actionable intelligence can change the fate of entire harvests, turning potential crises into manageable situations.

Data accumulation is a key enabler of effective AI-driven disease detection. By continually updating models with real-world data, the systems become smarter and more accurate over time. Farmers contribute to this vast repository of information simply by using these technologies, feeding back their unique field conditions, disease incidents, and treatment outcomes.

Additionally, farmer communities stand to benefit from these AI systems through shared platforms where data and insights are exchanged. Such platforms can function like a collective brain, where experiences from diverse regions and crop types converge, enriching the pool of knowledge and leading to more precise disease management strategies.

It's worth noting the role that satellite imagery plays in this ecosystem. Satellites equipped with multispectral and hyperspectral sensors capture images of fields from space, providing another layer of data. These images, with their varying wavelengths, offer critical information on crop vigor, plant stress, and potential disease onset. By analyzing this data, AI systems can monitor large swaths of agricultural land simultaneously, making disease detection scalable and efficient.

In implementing these technologies, collaboration and education are crucial. Partnerships with tech companies, agricultural research institutions, and local extension services foster the development of these AI tools and ensure they are accessible to farmers. Training programs and workshops geared towards farmers can demystify these technologies, encouraging wider adoption and enhancing farming practices.

Despite the significant promise that AI holds in disease detection and prevention, some challenges persist. Smallholder farmers in remote regions may lack the required infrastructure, internet connectivity, or financial resources to adopt these technologies. Addressing these gaps through targeted subsidies, affordable tech solutions, and initiatives focused on digital literacy can help bridge this divide.

AI's role in disease detection also has a broader societal impact. Reduced crop loss translates into more stable food supplies, less wastage, and improved food security, contributing to the overall well-being of populations. Moreover, healthier crops with fewer chemical treatments resonate with consumers looking for sustainable and safer food options.

The journey ahead involves continuous innovation and adaptation. As AI technologies evolve, so too will their applications in agriculture. Imagine a future where biosensors in the fields communicate with AI systems in real time, creating a dynamic and responsive farming environment. This synergy between man and machine promises a new era of disease-free crops and bountiful harvests.

In this landscape, AI isn't just a tool; it's an ally in the relentless pursuit of agricultural excellence. By integrating data analytics, machine learning, and real-world farming practices, AI systems are setting new standards for disease detection and prevention, driving the agricultural sector towards a more sustainable and prosperous future.

To maintain momentum, stakeholders across the agriculture value chain must engage in dialogue and cooperation. Governments can play a supportive role by crafting policies that encourage AI adoption and provide a framework for ethical considerations around data privacy. Meanwhile, tech companies can continue innovating, developing user-friendly interfaces and accessible solutions tailored to diverse farming needs.

The narrative unfolding in disease detection and prevention is compelling and full of promise. It's a testament to how far agricultural technology has come and an inspiring glimpse of what lies ahead. Healthier crops lead to healthier communities, paving the way for a future where food security is no longer a challenge, but a guaranteed reality across the globe.

Chapter 5:
Soil Analysis and Health

The heart of any agricultural venture lies beneath our feet— the soil. Understanding its composition and health is crucial for optimizing crop yields and ensuring sustainable farming practices. AI-driven soil analysis not only automates the traditionally labor-intensive process of soil testing but also elevates it with unprecedented precision. Advanced algorithms can analyze extensive datasets to offer real-time insights on soil pH, nutrient levels, and moisture content. This enables farmers to make informed decisions about fertilization and irrigation, tailoring interventions to soil-specific needs. By leveraging machine learning models, we can predict how soil properties will change over time in response to agricultural activities, allowing for proactive measures that mitigate degradation and promote long-term fertility. Ultimately, AI empowers farmers to cultivate healthier soils, leading to more resilient crops and contributing to the overarching goal of sustainable agriculture.

AI-driven Soil Testing

Soil testing has long been a cornerstone of successful farming, offering insights into soil composition, nutrient levels, and other critical factors that influence crop health and yield. However, traditional soil testing methods can be time-consuming, labor-intensive, and occasionally imprecise. This is where artificial intelligence (AI) steps in to revolutionize soil analysis, bringing unprecedented accuracy, efficiency, and actionable insights to the table. AI-driven soil testing not only speeds up the process but also enables farmers to make data-driven decisions that optimize their soil health and overall crop performance.

At its core, AI-driven soil testing employs various technologies, including machine learning algorithms, computer vision, and sensor data integration, to analyze soil samples more efficiently than ever before. By collecting vast amounts of data from soil sensors, drones, and satellite images, AI systems can offer a comprehensive understanding of soil conditions across different fields and regions. As a result, farmers can identify nutrient deficiencies, soil pH imbalances, and other critical issues without delay, enabling timely interventions and targeted treatments.

One of the most significant advantages of AI in soil testing is its ability to continually learn and improve from the data it analyzes. Machine learning algorithms can process terabytes of information and recognize patterns that would be imperceptible to the human eye. With every new data point, the AI system becomes better at predicting soil health trends and offering more accurate recommendations. This constant learning cycle ensures that the insights provided remain relevant and highly accurate, benefiting farmers throughout the growing season.

Imagine a field outfitted with an array of soil sensors, each transmitting data in real-time to a central AI system. This system continuously monitors soil moisture levels, nutrient concentrations, and even the presence of pests or pathogens. As conditions change, the AI can alert farmers to issues that need immediate attention, such as excessive soil acidity or a sudden drop in essential nutrients like nitrogen or phosphorus. Armed with this information, farmers can apply the right fertilizers or soil amendments, reducing waste and improving crop yields.

Moreover, AI-driven soil testing can integrate historical data to offer long-term insights into soil health. By analyzing years of soil data alongside weather patterns and farming practices, AI systems can predict future soil health trends and suggest proactive measures. This foresight can be invaluable for farmers looking to implement sustainable farming practices and ensure the longevity of their land. By addressing potential problems before they become critical, farmers can maintain soil fertility and health, leading to more robust and resilient crops.

In addition to enhancing soil health, AI-driven soil testing also plays a significant role in environmental conservation. Traditional soil testing methods often involve the use of chemicals and other potentially harmful substances, which can contribute to soil degradation and pollution. AI's non-invasive and data-driven approach minimizes environmental impact, promoting more sustainable and eco-friendly farming practices. Additionally, by optimizing fertilizer and pesticide usage, AI-driven soil testing helps reduce runoff and pollution, contributing to the overall health of surrounding ecosystems.

A practical example of this technology in action can be found in precision farming, where AI-driven soil testing works hand in hand with other AI-based systems to optimize every aspect of crop production. Farmers can use soil health data to create precise planting maps, ensuring that each seed is sown in optimal conditions. During the growing season, AI systems can monitor the soil and make real-time adjustments to irrigation, fertilization, and pest control strategies, maximizing crop health and yield. This holistic approach not only boosts productivity but also conserves resources and reduces environmental impact.

Moreover, the integration of AI in soil testing democratizes access to advanced agricultural insights. Small-scale and resource-limited farmers can benefit from cloud-based AI platforms that offer soil analysis services without the need for expensive on-site equipment. Using a smartphone or a simple sensor kit, farmers can collect soil samples and upload data to an AI platform, receiving detailed analysis and recommendations in return. This accessibility empowers more farmers to adopt data-driven practices, leveling the playing field and contributing to global food security.

Machine learning algorithms have also opened up new frontiers in predictive soil management. By leveraging predictive analytics, farmers can anticipate changes in soil conditions based on various factors such as crop type, weather forecasts, and historical data. This predictive capability enables farmers to plan more effectively, ensuring that they are always prepared for potential soil health issues before they impact crop growth. It also allows for the fine-tuning of crop rotations and other long-term strategies that enhance soil fertility and sustainability.

AI-driven soil testing is not just a tool; it's a paradigm shift in how we approach soil health and farm management. Its ability to analyze complex data sets and offer actionable insights in real-time sets it apart from traditional methods. By embracing this technology, farmers can transform their practices, leading to healthier soils, more productive fields, and a more sustainable future for agriculture. This transformation is not just about increasing yields or profits; it's about cultivating a deeper understanding of the dynamic relationship between soil, crops, and the environment, fostering a more holistic and mindful approach to farming.

In conclusion, AI-driven soil testing represents a monumental leap forward in agricultural technology. Its combination of efficiency, accuracy, and sustainability makes it an invaluable asset for modern farmers striving to meet the challenges of feeding a growing global population. By integrating cutting-edge AI technologies with traditional farming practices, we can unlock the full potential of our soils, ensuring that they remain fertile and productive for generations to come. This fusion of technology and agriculture heralds a new green revolution, one that promises to not only enhance food production but also safeguard our planet's precious resources.

As we continue to explore and innovate at the intersection of AI and agriculture, the possibilities for improving soil health and overall farm productivity are endless. The future of farming is data-driven, and AI-driven soil testing lies at the heart of this transformation. By harnessing the power of AI, we can cultivate a world where agriculture is more efficient, sustainable, and resilient, paving the way for a brighter and more bountiful future.

Optimizing Soil Health for Better Yields

Soil health forms the bedrock of agricultural success. Indeed, plants rely on this rich matrix not just for physical support, but also for critical nutrients and water that influence their growth and productivity. Soil optimization is thus not only practical but essential if we are to achieve better yields in sustainable ways. As we dig deeper into the intersection of soil health and AI-driven methodologies, it becomes clear that modern technology holds the promise to revolutionize how we perceive and enhance our soil's potential.

At its core, soil is more than just dirt. It's a dynamic and intricate system teeming with life. The diverse microorganisms within it – bacteria, fungi, and other tiny creatures – play a vital role in nutrient cycling, organic matter decomposition, and even disease suppression. However, the health of this system can be hampered by several factors, be it unsustainable farming practices, climate changes, or the overuse of chemical fertilizers. Understanding and addressing these issues is fundamental for ensuring long-term agricultural productivity and environmental sustainability.

Artificial Intelligence (AI) has emerged as a pivotal tool in optimizing soil health. By leveraging the enormous volumes of data generated from various sources such as satellite imagery, soil sensors, and historical crop data, AI systems can provide actionable insights that help in making informed decisions. For instance, AI-driven soil testing can now analyze soil composition with unprecedented accuracy, identifying deficiencies or excesses in vital nutrients and recommending precise types and amounts of fertilizers needed. This level of precision was unimaginable in traditional soil analysis and ensures optimum utilization of resources while minimizing environmental impact.

One of the remarkable AI applications in soil health is predictive modeling. By utilizing historical data and machine learning algorithms, these models can forecast potential soil health issues before they even occur. These predictions can range from nutrient depletion risks to the probable impacts of upcoming weather conditions on soil moisture levels. Armed with this foresight, farmers can undertake preventive measures well in advance, thereby safeguarding their crops and ensuring sustained yields.

Moreover, AI doesn't just stop at predictive capabilities. It also facilitates continual monitoring through autonomous soil sensors and drones equipped with advanced imaging technologies. These tools provide real-time data on various soil parameters such as moisture content, pH levels, and organic matter. Not only does this mean less guesswork, but it also means interventions can be more timely and highly targeted. Consider the potential water savings when irrigation systems are aligned with real-time soil moisture data provided by AI. Such harmonization ensures that plants receive just the right amount of water they need, reducing wastage and conserving this precious resource.

In terms of restocking the soil's natural nutrients, AI can enable a more balanced approach toward organic and inorganic inputs. Through precision agriculture, AI can determine the most effective organic matter to introduce, be it crop residues, compost, or green manures. These materials not only enrich the soil but also enhance its structure, water retention capacity, and biological activity. By reducing dependency on chemical fertilizers, this method promotes a more sustainable and resilient farming ecosystem.

However, the optimization of soil health through AI isn't solely about technology. It also involves a paradigm shift in the farmer's mindset. Traditional farming practices that heavily relied on experience and intuition now need to be integrated with data-driven insights. This hybrid approach can sometimes be challenging, requiring education and ongoing support. Farmers must be trained not only to understand the data but more importantly, to trust and act upon it. Successfully navigating this transition means fostering a culture that values continuous learning and technological adaptation.

For instance, collaborative platforms that facilitate knowledge exchange between scientists, agronomists, and farmers can be instrumental. These communities can share success stories, troubleshoot issues, and discuss the latest advancements in AI applications for soil health. Harnessing such collective intelligence accelerates the adoption of innovative practices, ensuring that the benefits of AI in soil health optimization aren't confined to a select few but become widespread.

A critical aspect of soil health often overlooked is the cumulative impact of farming practices. AI can play a pivotal role in assessing long-term soil health trends, helping in the formulation of strategies aimed at sustainability. For example, crop rotation guided by AI can suggest optimal planting sequences that naturally replenish soil nutrients, mitigating the risk of soil fatigue. Similarly, AI can advise on the best cover crops to grow during off-seasons, preventing soil erosion and maintaining its fertility.

Moreover, AI's contribution extends to addressing the broader environmental implications of farming. For instance, carbon sequestration – an important process for mitigating climate change – can be significantly enhanced through healthy soil. AI systems can identify the most suitable regions and farming practices for carbon capture, enabling farmers to participate in carbon credit programs, thus providing both ecological and economic benefits.

As precision agriculture continues to evolve, AI's role in optimizing soil health will undoubtedly expand. The future points to even more integrated systems where AI interacts seamlessly with IoT devices, blockchain for secure data transactions, and even augmented reality to provide farmers with intuitive interfaces. Such advancements will only enhance the efficacy of current practices and open new avenues for exploring soil health.

In conclusion, optimizing soil health through AI is not just a necessity but an opportunity to revolutionize agriculture. The dynamic interplay of cutting-edge technologies and time-tested agricultural wisdom holds the key to unlocking unprecedented levels of productivity and sustainability. By continually striving to understand and enhance the richness of our soil, we pave the way for a future where food security and environmental preservation go hand in hand. The soil beneath our feet is, in essence, a living entity, and with AI as our ally, nurturing it promises bountiful rewards for generations to come.

Chapter 6:
Water Management

Water management is a linchpin of sustainable agriculture, especially as climate variability increases unpredictability in water availability. AI-powered solutions offer a transformational approach to irrigation and water usage, enabling farmers to optimize resources down to individual plants. By leveraging advanced sensors and real-time analytics, AI systems can monitor soil moisture levels and adjust irrigation schedules with pinpoint accuracy, ensuring each crop gets exactly what it needs without waste. This not only conserves water but also enhances crop yield and quality. Such precision would be unattainable through traditional methods, making AI indispensable in modern water management strategies. The fusion of technology and agriculture here signals a significant movement towards environmentally responsible farming practices, driving productivity while conserving our planet's most vital resource.

AI Solutions for Efficient Irrigation

In the expansive realm of water management, irrigation stands as one of the most critical yet challenging aspects. Efficient irrigation isn't just about pouring water onto fields; it's an intricate balance of delivering just the right amount of water at the right time to ensure optimal crop growth while conserving resources. This is where artificial intelligence (AI) steps in as a transformative force, offering innovative solutions to age-old problems and redefining the future of irrigation.

AI-driven irrigation solutions utilize a combination of machine learning algorithms, real-time data analytics, and sensor technologies to dynamically adjust water delivery to crops. Unlike traditional irrigation systems that often operate on fixed schedules, AI-based systems continually learn and adapt to changing conditions. These systems analyze a vast array of data, including weather patterns, soil moisture levels, crop health metrics, and more, to make informed and responsive irrigation decisions.

One of the standout benefits of using AI in irrigation is **precision**. Farmers can move away from the one-size-fits-all approach, tailoring irrigation to the specific needs of different crops and even individual plants. Sensors embedded in the soil gather real-time data on moisture levels, which is processed by AI algorithms to determine the exact amount of water required. This precision not only enhances crop yields but also significantly reduces water wastage.

A critical aspect of AI-enhanced irrigation is predictive analytics. These systems can forecast future water needs based on historical data and predictive weather models. By anticipating dry spells or periods of heavy rainfall, AI can proactively adjust irrigation schedules, ensuring crops remain adequately hydrated without the risk of overwatering or underwatering. This forward-thinking approach leads to optimal resource utilization and enhances crop resilience against climatic variations.

The integration of AI in irrigation practices also fosters sustainability. By optimizing water usage, these systems help to preserve vital water resources, which is especially crucial in regions plagued by water scarcity. In drought-prone areas, the ability to predict and efficiently manage water use can be the difference between a bountiful harvest and a failed one. Sustainable water management through AI not only benefits individual farmers but also contributes to broader environmental conservation efforts.

Moreover, AI solutions for efficient irrigation often come equipped with remote monitoring capabilities. Farmers can manage and monitor their irrigation systems from anywhere using mobile apps or web interfaces. This remote oversight allows for immediate adjustments to irrigation schedules or responding to sensor alerts, thereby minimizing the risk of crop damage due to water stress. This level of control and flexibility can be a game-changer, especially for large-scale farms spanning extensive geographical areas.

AI doesn't just help in real-time adjustments and predictive planning; it also facilitates long-term water management strategies. By continually analyzing data over multiple growing seasons, AI can provide insights into trends and patterns that might not be immediately evident. Farmers can then develop more comprehensive irrigation plans, considering both short-term needs and long-term sustainability goals. Over time, this adaptive learning mechanism ensures that irrigation practices are continually refined and improved.

Another exciting development in AI-fueled irrigation is the integration of drone technology. Equipped with advanced sensors and cameras, drones can survey vast tracts of farmland, capturing high-resolution images and data points. When processed by AI systems, this data can reveal discrepancies in water distribution, identify areas of potential stress, and even detect irrigation system malfunctions. This bird's-eye view of irrigation efficiency is a powerful tool for precision agriculture, driving even more targeted and effective water management practices.

For small-scale farmers and those in developing regions, AI-powered irrigation solutions can be particularly impactful. These farmers often face significant challenges in accessing advanced technologies and resources. AI can democratize access to efficient irrigation techniques through affordable, scalable technologies. Smartphone apps and low-cost sensors can provide critical data and insights, enabling even the smallest farms to benefit from precision irrigation practices.

Investment in AI irrigation technologies is not just an economic imperative; it's a moral one. Water is a finite and precious resource, and agriculture is one of its largest consumers. By adopting AI solutions, the agricultural sector can lead in responsible water usage, setting a precedent for other industries to follow. Furthermore, food security is closely linked to water management. Efficient irrigation powered by AI can ensure consistent crop production, reducing the vulnerability of global food supplies to climatic and environmental disruptions.

Partnerships and collaborations play a crucial role in advancing AI irrigation technologies. Governments, academic institutions, and private enterprises must work together to develop, test, and deploy these technologies on a broader scale. Such collaborative efforts can drive innovation, lower costs, and enhance the accessibility of AI-driven irrigation solutions, ensuring that their benefits are widely realized.

As we look towards the future, the potential of AI in irrigation continues to expand. Emerging technologies like Internet of Things (IoT) devices, advanced machine learning models, and better computational resources are continuously enhancing the capabilities of AI systems. The future may see fully autonomous irrigation systems that require minimal human intervention, allowing farmers to focus on other critical aspects of farming.

The journey towards efficient irrigation through AI is not without its challenges. Initial setup costs, ongoing maintenance, and the need for technical expertise can be barriers to adoption. However, the long-term benefits of water savings, increased yields, and environmental sustainability make it a worthwhile investment. Support from agricultural extension services, government subsidies, and educational initiatives can help alleviate these challenges, paving the way for widespread adoption of AI in irrigation.

In conclusion, AI solutions for efficient irrigation stand at the forefront of modern agriculture. By leveraging technology to make smarter, more informed decisions, farmers can achieve sustainable water management and enhance their crop productivity. This convergence of technology and agriculture not only addresses immediate challenges but also lays the foundation for a more resilient and resource-efficient future. As we harness the power of AI, we move closer to a world where agriculture thrives while preserving our precious natural resources.

Monitoring Water Usage and Soil Moisture

Water scarcity is one of the most pressing issues modern agriculture faces. Proper water management not only ensures high-yielding crops but also sustains the environment. Monitoring water usage and soil moisture in real-time allows farmers to make data-driven decisions, optimizing every drop of water used. The integration of AI in this process is pivotal, as it helps in precise and efficient water management, enabling sustainable farming practices.

A critical aspect of water management is understanding the soil moisture levels. Traditional methods of gauging soil moisture, like manual sampling, are not only labor-intensive but also often inaccurate. AI-based solutions have revolutionized the accuracy and efficiency of monitoring these levels. Smart sensors deployed in the fields constantly measure parameters like soil moisture, temperature, and electrical conductivity. This data is fed into AI algorithms that analyze and predict soil moisture levels, guiding irrigation systems to deliver the right amount of water at the right time.

For instance, machine learning models trained on historical and real-time data can predict periods of water stress in crops. These models analyze factors such as weather forecasts, soil type, and crop water demand. By integrating weather data, AI systems can postpone irrigation when rain is predicted, conserving water resources. Similarly, during dry spells, these systems can ensure crops receive the necessary watering, thus mitigating the risk of water stress.

AI-driven monitoring systems make use of various data sources, from satellite imagery to IoT devices placed in fields. Satellite imagery provides a macro view of the moisture levels across the entire farming landscape. This, combined with IoT sensors, offers a granular, real-time view of soil conditions. Algorithms process this wealth of data to provide actionable insights, ensuring that water resources are utilized optimally.

One of the most significant benefits of AI in water management is its predictive capabilities. By analyzing past data trends, AI can forecast future water needs. Algorithms can model complex relationships between weather conditions, soil moisture levels, and crop requirements. Suppose a particular pattern indicates that a critical moisture level will be reached soon. In that case, the system can alert the farmer or automatically adjust irrigation schedules.

Drip irrigation and smart irrigation systems, when combined with AI, transform the efficiency of water usage. Drip irrigation delivers water directly to the plant roots, minimizing evaporation and runoff. When controlled by AI systems, drip irrigation can be fine-tuned to supply just the right amount of water, reducing waste and ensuring optimal soil moisture levels. This precise control is especially crucial in regions facing severe water scarcity, making every drop count.

In places with uneven rainfall distribution, AI helps balance the unpredictability. By analyzing historical weather and crop yield data, AI can develop customized irrigation schedules for different crop types and growth stages. This approach not only saves water but also boosts crop health and yield, making farming more sustainable and resilient to climate change impacts.

Furthermore, AI aids in detecting potential issues such as irrigation system leaks or blockages. If a sensor detects unusual moisture patterns or a sudden drop in water pressure, the AI system can alert farmers so they can take remedial actions promptly. Early detection of such issues prevents water wastage and ensures that crops receive uniform hydration.

One remarkable example of this technology in action is the application of AI in vineyards. Growers use AI-powered systems to monitor soil moisture through sensors and aerial drones. These systems ensure that vines get the optimal amount of water, thus enhancing grape quality and yield. The AI models can even account for the microclimates within the vineyard, providing nuanced irrigation strategies that traditional methods would miss.

The integration of AI also opens doors for farmers to participate in water conservation programs. Governments and environmental organizations often encourage sustainable practices through incentives. AI can help farmers demonstrate their efficient water use, making them eligible for such programs. This not only benefits the environment but also provides an economic advantage to farmers.

Another facet worth mentioning is the role of AI in educating farmers about water management. User-friendly platforms powered by AI can offer insights, recommendations, and even tutorials on optimizing water usage. These platforms often come with mobile applications, providing farmers with real-time updates and alerts. Farmers can interact with these systems to get advice tailored to their specific conditions, thus bridging the gap between traditional farming practices and modern technology.

AI also supports community-level water management initiatives. For instance, in regions where multiple farms share a common water source, coordinated water usage becomes essential. AI systems can optimize water distribution among farms, ensuring equitable use and avoiding conflicts. This collaborative approach promotes a sense of community and collective responsibility towards water conservation.

Despite the clear benefits, the adoption of AI in monitoring water usage and soil moisture faces some challenges. The initial cost of setting up AI systems can be prohibitive for small-scale farmers. However, as the technology matures, costs are expected to decrease, making it more accessible. Initiatives by governments and NGOs to subsidize these technologies can also play a crucial role in widespread adoption.

Data privacy is another concern. Farmers may be hesitant to adopt AI systems fearing data misuse. Ensuring robust data security and transparent data usage policies is vital for gaining farmers' trust. Moreover, educating farmers about the long-term benefits of AI can help alleviate their concerns, promoting wider acceptance of the technology.

In conclusion, AI's role in monitoring water usage and soil moisture is transformative. It empowers farmers with precise, data-driven insights, fostering efficient water use practices. By predicting future water needs and optimizing irrigation schedules, AI contributes significantly to sustainable agriculture. The journey of adopting AI in water management might have its hurdles, but the potential rewards in terms of water conservation, improved crop yield, and economic benefits make it a worthwhile endeavor. As we continue to innovate and refine these technologies, the dream of achieving a sustainable and resilient agricultural ecosystem comes closer to reality.

Chapter 7:
Weed and Pest Control

Imagine a field where weeds and pests are instantly identified and managed with pinpoint accuracy, thanks to the marvels of AI. This isn't speculative fiction; it's the new reality transforming agriculture. By leveraging advanced machine learning algorithms, AI systems can differentiate between crops and invasive species in real time, allowing for the precise application of herbicides and pesticides. What's more, predictive analytics can forecast pest outbreaks, enabling proactive measures rather than reactive responses. Picture drones zipping through fields, scanning for early signs of trouble, and sending data back to AI models that recommend targeted interventions. This not only reduces the overuse of chemicals but also minimizes crop loss, leading to more sustainable farming practices. Indeed, with AI on our side, we're moving from guesswork to data-driven weed and pest control, ensuring healthier crops and better yields.

Identifying and Managing Weeds with AI

Weeds have long been the bane of farmers' existences, siphoning nutrients, water, and sunlight away from valuable crops. Traditionally, managing these uninvited guests has required a combination of physical labor and chemical herbicides, both of which come with their own sets of challenges and drawbacks. Enter artificial intelligence, a game-changer in the world of agriculture, bringing precision and efficiency to weed identification and management.

AI leverages machine learning algorithms and image recognition technologies to identify weeds accurately in a field. Advanced cameras attached to drones or autonomous ground vehicles scan crops, capturing high-resolution images that are then analyzed by AI systems. These systems can differentiate between crop plants and various weed species with a level of accuracy that is difficult for human eyes to achieve, especially over large areas.

The magic of AI lies in its ability to learn and adapt. Machine learning algorithms are trained on vast datasets of images that include both crops and weeds at various growth stages. Over time, these algorithms get better at identifying even the most inconspicuous weeds. Furthermore, an AI system can be trained to recognize different types of weeds that might be specific to a region or crop, providing a tailored solution for farmers.

Real-time data processing is another significant advantage. In traditional methods, by the time weeds are identified and treated, they might have already affected the crop yield. AI systems can process the captured images almost instantaneously, allowing for immediate action. This real-time capability ensures that farmers can tackle weed problems before they escalate, effectively nipping them in the bud.

One of the standout features of AI-powered weed management is its precision. Unlike broad-spectrum herbicide application, which can harm the environment and affect the health of soil and plants, AI enables precision spraying. Autonomous machinery guided by AI can target individual weeds or small clusters, spraying herbicides only where needed. This targeted approach minimizes chemical usage, reduces costs, and promotes a healthier ecosystem.

The benefits go beyond just weed control. By optimizing the use of herbicides, AI helps in maintaining soil health, which is crucial for long-term agricultural sustainability. Overuse of chemicals can lead to soil degradation and decreased fertility, but precision spraying mitigates this risk. Additionally, the reduced chemical runoff helps protect nearby water bodies, preserving aquatic life and ensuring cleaner water resources.

Financially, integrating AI into weed management can be a boon for farmers. The initial investment in AI technology and autonomous machinery may seem steep, but the long-term savings on labor and chemicals, as well as the improved crop yields, make it a worthy investment. Financial models and case studies show that farms using AI for weed management often see a significant return on investment within just a few seasons.

It is also worth noting the role of AI in promoting sustainable farming practices. The reduced need for chemical inputs aligns with organic farming principles and can help farmers transition to more sustainable practices without compromising on productivity. AI-powered weed management systems can be integrated with other eco-friendly practices like crop rotation and cover cropping, creating a holistic approach to sustainable agriculture.

Moreover, the data gathered through AI systems for weed identification can be invaluable for research and development. Detailed records of weed types, distribution patterns, and growth cycles can provide insights that help in breeding more resilient crop varieties and developing new weed management strategies. This data-centric approach opens up new avenues for innovation in agricultural science.

Farmers can also benefit from the scalability of AI solutions. Whether managing a small plot or a sprawling farm, AI systems can be adapted to fit the scale of operations. Drones equipped with AI can cover large fields quickly, while smaller, autonomous machines can navigate tighter, more intricate farm layouts. This flexibility ensures that AI-driven weed management is accessible to a broad spectrum of agricultural enterprises.

There are several success stories that illustrate the effectiveness of AI in weed management. Farms in North America, Europe, and Australia have reported significant improvements in weed control efficiency and crop yields after adopting AI technologies. These real-world examples serve as a testament to the potential of AI to revolutionize weed management and, by extension, farming as a whole.

Of course, no technology is without its challenges. The integration of AI into weed management systems requires training and expertise. Farmers need to be familiar with the operation of drones, autonomous vehicles, and the underlying AI software. However, ongoing advancements in user interfaces and support systems are making these technologies more user-friendly, even for those who may not be tech-savvy.

Another challenge lies in the initial cost and maintenance of AI systems. While the long-term benefits are clear, the upfront costs can be a barrier for small-scale farmers. However, with increasing interest from both public and private sectors, various funding and subsidy options are becoming available to help farmers adopt these innovative technologies.

Looking ahead, the future of weed management with AI is promising. Next-generation AI systems are likely to incorporate even more advanced features such as predictive analytics, which can forecast weed infestations based on historical data and environmental conditions. This proactive approach will allow farmers to implement preemptive measures, further enhancing the efficiency of weed control strategies.

Furthermore, AI-driven systems will continue to evolve, incorporating more sensors and data sources to refine their accuracy. Integration with other smart farming technologies, such as soil moisture sensors and climate monitoring tools, will enable a more comprehensive approach to farm management. This interconnected ecosystem of AI tools will empower farmers to optimize all aspects of their operations, driving the next green revolution.

In summary, the application of AI in weed identification and management stands as a testament to the transformative power of technology in agriculture. By bringing precision, efficiency, and sustainability to weed control, AI is helping to create a future where farming is both more productive and environmentally friendly. As we continue to innovate and refine these technologies, the possibilities for their impact on agriculture are bound to grow, ensuring that we can meet the food demands of an ever-increasing global population. The integration of AI into weed management is not just a leap forward in farming; it's a giant step towards a more sustainable and prosperous future.

AI for Pest Prediction and Management

In the evolving world of agriculture, accurate pest prediction and effective management have always been critical. Traditional methods, while useful, often fall short in providing the precision, scale, and timeliness required to keep up with today's agricultural demands. This is where artificial intelligence (AI) comes into play, offering transformative solutions that bring both efficiency and sustainability to pest control.

Gone are the days when farmers relied solely on manual inspections and historical data to anticipate pest outbreaks. With the advent of AI, we can now leverage complex algorithms and machine learning models to predict pest patterns with remarkable accuracy. For instance, AI systems can analyze vast amounts of data collected from various sources—satellite imagery, weather forecasts, soil conditions, and even social media chatter—to forecast pest invasions before they become problematic.

Imagine an AI-powered system that alerts farmers about potential pest outbreaks days, or even weeks, in advance. This proactive approach allows them to take preemptive measures, such as applying targeted pesticides or deploying natural predators, thereby reducing crop damage and enhancing yields. Moreover, the precision offered by AI ensures that pest control measures are applied only where necessary, significantly cutting down on chemical use and promoting environmental sustainability.

Let's delve deeper into the mechanics of how AI accomplishes this. Machine learning models, specifically trained on historical pest data, can identify patterns and correlations that are invisible to the human eye. These models continuously learn and adapt as they are fed new data, becoming increasingly accurate over time. Factors like temperature, humidity, crop type, and growth stage are all considered, creating a multifaceted approach to pest prediction.

Moreover, AI-backed systems can incorporate data from remote sensing technologies and Internet of Things (IoT) devices installed on farms. These devices collect real-time data on environmental conditions, crop health, and pest movements. By integrating this real-time data with predictive models, AI offers unparalleled insights and recommendations that are grounded in the current state of the farm, not just historical trends.

One practical implementation of AI in pest management is the use of drone technology. Equipped with high-resolution cameras and sensors, drones can survey large tracts of land quickly and efficiently, capturing detailed images that reveal the early signs of pest infestations. The captured data is then processed through AI algorithms that identify pest activity with astonishing precision, allowing for immediate action.

The scalability of these AI solutions can't be overstated. Whether you're managing a sprawling commercial farm or a small family-owned plot, AI can be scaled to meet your specific needs. Small-scale farmers, who often lack the resources for extensive pest control measures, stand to benefit immensely. AI-driven pest management tools are often accessible and can transform the efficiency and productivity of smaller operations.

Field trials and existing case studies have demonstrated the efficacy of AI in pest prediction and management. For example, a specific AI model predicted locust swarms in East Africa, allowing local farmers to take preventive measures and save their crops. In another instance, vineyards in California utilized AI to identify the early stages of grapevine moth infestations, dramatically reducing crop loss and improving grape quality.

Beyond the technology itself, what makes AI truly revolutionary in pest management is its potential for continuous improvement. Machine learning models can constantly adapt to new data, refining their predictions and recommendations over time. This dynamic nature ensures that AI solutions remain relevant and effective even as pest behaviors and environmental conditions evolve.

However, the integration of AI into pest management isn't without its challenges. Data quality and quantity are paramount— poor or insufficient data can significantly impact the accuracy of predictions. Therefore, ensuring robust data collection mechanisms and training models with diverse datasets is crucial. Additionally, there might be initial resistance from farmers accustomed to traditional methods. Overcoming this requires education and training to demonstrate the tangible benefits of AI adoption.

Ethical considerations also play a role, especially when it comes to data privacy and the use of automated pest control methods. Ensuring that AI systems operate transparently and that farmers are aware of how their data is being used can foster greater trust and acceptance. Collaborative efforts between technologists, agronomists, and farmers are essential to create solutions that are not only effective but also equitable and ethical.

Looking ahead, the future of AI in pest prediction and management holds exciting possibilities. Continued advancements in AI and machine learning, coupled with improvements in sensor technologies and data analytics, will further enhance the precision and reliability of these systems. Integrating AI with other emerging technologies, such as blockchain for data validation and augmented reality for field inspections, could create comprehensive pest management ecosystems that are robust and resilient.

As we push the boundaries of what's possible with AI, the ultimate goal remains clear: to create a sustainable agricultural system that minimizes pest-related losses, reduces dependency on chemical interventions, and promotes healthy crop growth. The journey is challenging but attainable, and the transformative potential of AI in pest prediction and management brings us one step closer to realizing this vision.

Chapter 8:
Autonomous Machinery

The agricultural landscape is being revolutionized by the advent of autonomous machinery, transforming age-old practices and significantly elevating productivity. AI-powered tractors and harvesters now move effortlessly through fields, executing tasks with precision that human hands can't match. These machines are not only capable of navigating diverse terrains but also optimizing their routes to conserve energy and reduce operational costs. Robots are stepping into roles traditionally held by farmworkers, like planting seeds and maintaining crops, ensuring consistency and efficiency. With sensors and machine learning algorithms, these autonomous entities can adapt to real-time conditions, making split-second decisions that can determine yield success. The fusion of artificial intelligence and robotics is thus paving the way for a new era where farming is more efficient, sustainable, and scalable, heralding what could be the next green revolution.

AI-powered Tractors and Harvesters

Innovation in farming technology has taken a giant leap forward with the advent of AI-powered tractors and harvesters. These autonomous machines are not only revolutionizing traditional agricultural practices but also bringing a new wave of efficiency, precision, and sustainability. With their ability to perform tasks that once required extensive human labor, AI-enhanced tractors and harvesters are setting the stage for a modern agricultural renaissance.

At the heart of these intelligent machines lies a complex network of sensors, cameras, and machine learning algorithms. These components work in tandem to provide tractors and harvesters with the capability to make real-time decisions based on a plethora of data points. For instance, AI algorithms can analyze soil conditions, crop health, and weather patterns to optimize planting, fertilizing, and harvesting schedules. The result is a process that's more efficient and yields a higher output, thereby maximizing both time and resources.

Traditional tractors have evolved from manually operated machines into sophisticated, self-driving units. Equipped with GPS systems, AI-powered tractors can navigate fields with pinpoint accuracy. This precision ensures that seeds are planted at the optimal depth and spacing, reducing waste and promoting healthier plant growth. Additionally, AI-driven tractors can adapt to varying field conditions, making real-time adjustments to their operations.

Harvester technology has also seen significant advancements. Modern harvesters can identify the ripeness of crops and make harvesting decisions on the fly, ensuring that only the best-quality produce is collected. This capability reduces the loss of yield due to premature or late harvesting. Some AI-powered harvesters are even equipped with robotic arms that can handle delicate crops, minimizing damage and improving overall quality.

Beyond the basic functionalities, these machines are designed to handle extensive data analytics. They can monitor crop performance throughout the growing season and provide farmers with actionable insights. This level of data-driven decision-making helps farmers predict potential issues and take preemptive measures to mitigate risks. The integration of AI allows for a level of precision that manual methods simply cannot achieve.

Consider, for example, a scenario where a harvester is equipped with computer vision technology. This capability allows the machine to scan each crop, analyze its condition, and determine the best course of action. If a section of the field shows signs of disease, the harvester can alert the farmer and even apply targeted treatments autonomously. This targeted approach not only saves time and resources but also ensures the healthy growth of the remaining crops.

The autonomous nature of these machines also contributes to labor savings, addressing one of the significant challenges in modern agriculture: labor shortages. In many parts of the world, finding skilled labor for farming operations is becoming increasingly difficult. AI-powered tractors and harvesters can operate around the clock without fatigue, allowing for continuous production and timely completion of tasks. This is particularly beneficial during peak farming seasons when time is of the essence.

Environmental sustainability is another substantial benefit of AI-powered machinery. Precision farming techniques enabled by these machines lead to more efficient use of fertilizers, pesticides, and water. By applying these inputs only where and when needed, farmers can minimize their environmental footprint. Moreover, the data collected by these machines can help farmers implement more sustainable farming practices over time, contributing to long-term ecological balance.

Integration with other smart farming technologies amplifies the impact of AI-powered tractors and harvesters. For instance, data from drones and satellite imaging can be fed into the systems of these machines to enhance their decision-making capabilities. This integration creates a cohesive ecosystem where each component works synergistically to improve overall farm productivity and sustainability.

It's also worth noting the role of AI-powered machinery in overcoming challenges posed by climate change. Variability in weather patterns can severely impact farming operations. However, the predictive capabilities of AI allow farmers to make informed decisions by considering long-term weather forecasts and historical data. This foresight enables better planning and resilience against adverse weather conditions.

The economics of adopting AI-powered tractors and harvesters are compelling as well. While the initial investment may be significant, the long-term savings and productivity gains can offset these costs. Reduced labor expenses, efficient resource utilization, and higher crop yields contribute to a favorable return on investment. In addition, the ability to produce more with less input aligns with the broader goals of sustainable and profitable agriculture.

Farmers are not the only ones to benefit from these advancements. The entire supply chain stands to gain from the efficiencies introduced by autonomous machinery. Higher quality and more consistent produce make their way to markets faster, reducing waste and ensuring better prices for farmers. Consumers, in turn, enjoy fresh and high-quality products, fostering a positive cycle of benefits throughout the agricultural value chain.

One of the goals of advancing AI in agriculture is to democratize access to cutting-edge technology. Efforts are being made to make these sophisticated machines affordable and accessible to farmers of all scales. This inclusivity ensures that smallholder farmers can also reap the benefits of AI, contributing to a more equitable and productive agricultural sector.

As we look towards the future, the potential of AI-powered tractors and harvesters seems boundless. Ongoing research aims to further enhance their capabilities, integrating more advanced sensors, robotic elements, and machine learning models. With each innovation, these machines become better equipped to tackle the diverse challenges of modern farming.

In summary, AI-powered tractors and harvesters represent a leap forward in agricultural technology. They bring unprecedented levels of efficiency, precision, and sustainability to farming operations. By leveraging advanced sensors, machine learning algorithms, and data integration, these autonomous machines are transforming the way we grow and harvest crops. The resulting benefits extend to farmers, consumers, and the environment, positioning AI-powered machinery as a cornerstone of the next agricultural revolution.

Robots in Planting and Maintenance

Autonomy in agriculture has catalyzed a transformative change, promising unprecedented efficiency and accuracy in field operations. Particularly, the use of robots in planting and maintenance has become a game-changer, fusing artificial intelligence with mechanical prowess to reshape traditional farming methods. These autonomous machines are not just a vision of the future; they are actively at work in fields around the globe, addressing labor shortages, enhancing productivity, and minimizing environmental impact.

The integration of robots into planting practices represents a leap towards precision farming. Traditional sowing methods, characterized by human labor, variability, and inefficiencies, are being replaced by robots that ensure seeds are placed at the optimal depth and spacing. This isn't merely about mechanical planting; it's about intelligent planting, where each seed's placement is dictated by data-driven decisions. Robots equipped with AI can analyze soil composition, moisture levels, and weather forecasts to determine the exact time and place to plant seeds, hence optimizing germination and growth rates.

Take the example of robotic seeders, which have gained traction in modern agriculture. These machines follow pre-programmed paths with precision down to the centimeter, ensuring consistent planting patterns. They can adjust their actions based on real-time data inputs, making them incredibly adaptive to changing field conditions. This adaptability is crucial as it means higher yields and less resource wastage. Moreover, the efficiency of these robots allows them to work around the clock, completing in hours what would take human workers days.

Maintenance in agriculture extends beyond the initial planting phase, covering a wide array of activities crucial for healthy crop growth. Robots are also making significant strides in this realm. Autonomous weeding robots, for instance, utilize computer vision and machine learning algorithms to identify and eliminate weeds without harming the crops. This selective weeding reduces the need for chemical herbicides, promoting a more sustainable approach to pest management. These robots can traverse fields independently, leveraging GPS and sensors to detect their location and navigate accurately.

Another remarkable development is the emergence of robotic scouts. These are essentially mobile surveillance units that monitor crop health, soil conditions, and the presence of pests. Equipped with various sensors and cameras, they can collect data continuously, providing farmers with a real-time overview of their field's condition. By catching issues early, robotic scouts help in taking preventive measures, subsequently reducing crop losses and increasing overall farm productivity.

The versatility of maintenance robots is further illustrated by their application in crop spraying and fertilization. Autonomous sprayers can distribute pesticides and fertilizers with pinpoint accuracy, tailoring the application rates to the specific needs of different crop sections. This precision ensures that only the necessary amount of chemicals is used, reducing runoff and environmental damage. The adaptability of these robots means they can handle different types of crops and field conditions, making them invaluable assets in diversified farming operations.

More advanced models employ hyper-spectral imaging and AI to identify nutrient deficiencies or disease symptoms long before they're visible to the human eye. By using such sophisticated technologies, these robots ensure that interventions are carried out precisely when and where needed. This early-stage intervention is critical for maintaining crop health and maximizing yields, demonstrating the profound impact robots have on modern farming.

Soil maintenance is yet another crucial aspect where autonomous machinery is making a substantial impact. Robots equipped with soil sensors can conduct detailed soil analyses, scanning for nutrient levels, pH balance, and moisture content. This information is invaluable for making informed decisions about soil amendments, irrigation schedules, and crop rotations. Such robots operate autonomously, meaning they can cover large tracts of land quickly and efficiently, providing comprehensive data that enables farmers to maintain soil health over time.

Field robots also play a significant role in crop harvesting, which is often seen as the culmination of a successful planting and maintenance cycle. Once the crops are ready, harvesting robots equipped with AI systems can identify, pick, and sort the produce with precision and care. These robots not only speed up the harvesting process but also reduce post-harvest losses by ensuring that crops are handled gently and correctly.

Beyond their operational benefits, robots in planting and maintenance carry profound environmental advantages. By optimizing resource utilization—whether it's water, fertilizers, or pesticides—these machines help lower the ecological footprint of farming. Their precision-driven approach means fewer chemicals enter the environment, promoting biodiversity and preserving soil health. In agrarian regions where water resources are dwindling, such efficiency is not just beneficial; it's imperative for sustainable agriculture.

Financially, the initial investment in robotic technology for planting and maintenance may seem steep, but the long-term savings are substantial. Automation reduces labor costs, minimizes waste, and maximizes yield. Farmers can see a return on investment through improved productivity and reduced input costs. Moreover, the consistent performance of robots ensures that seasonal labor shortages do not disrupt farming activities, providing a stable operational environment.

The evolution of these technologies is also fostering partnerships between tech developers, agricultural experts, and farmers. Such collaborations are essential for refining robotic systems and ensuring they meet the practical needs of farms. This synergy between innovation and practicality is driving the sector forward, creating machines that are not only advanced but also indispensable for modern farming.

As we look to the future, the role of robots in planting and maintenance is set to expand even further. With ongoing advancements in AI, machine learning, and robotics, these machines will become more intelligent, autonomous, and versatile. Future iterations might include self-learning capabilities, where robots adapt to new challenges without human intervention, or even swarm robotics, where multiple machines operate collaboratively in a field, mimicking a hive of activity.

In conclusion, robots in planting and maintenance epitomize the convergence of technology and agriculture. They embody the promise of a more efficient, sustainable, and profitable farming practice. As these autonomous systems continue to evolve, they will undoubtedly drive forward the next green revolution, transforming the way we grow and maintain our food supply.

Chapter 9:
Predictive Maintenance

Predictive maintenance in agriculture leverages AI to foresee potential machinery failures before they happen, minimizing costly downtime and extending equipment life. By utilizing sensors and data analytics, AI systems monitor equipment in real-time, identifying patterns and anomalies that suggest imminent issues. This proactive approach allows farmers to schedule maintenance more efficiently, ensuring their machinery operates at peak performance during critical periods. For the tech-savvy farmer, integrating predictive maintenance means fewer interruptions during planting and harvest, reduced repair costs, and a smoother, more productive farming season. Ultimately, predictive maintenance not only boosts operational continuity but also contributes to the sustainability and economic viability of modern agricultural practices.

AI for Equipment Maintenance and Repair

As the sun sets on manual methods of equipment maintenance, AI emerges as a catalyst, reshaping the landscape of modern agriculture. Equipment maintenance and repair have always been fundamental aspects of farming, demanding significant time, labor, and resources. Traditionally, routine maintenance schedules and reactive repairs were the norm. However, these methods often led to considerable downtime, unexpected costs, and operational inefficiencies. With the advent of AI, the paradigm has shifted dramatically, ushering in an era of precision and predictive maintenance.

Imagine a world where farm machinery tells you it's about to break down before it actually does. AI-driven maintenance systems achieve exactly that, leveraging machine learning algorithms to predict when equipment is likely to fail. This transformative capability stems from the integration of numerous sensors embedded in farm machinery. These sensors continuously collect data such as temperature, vibration, and operational performance metrics. The collected data is then fed into AI models that analyze patterns and identify anomalies indicative of potential equipment failures.

One of the primary benefits of AI for equipment maintenance is its predictive capability. Predictive maintenance shifts the focus from reactive to proactive management. Using historical and real-time data, AI algorithms can forecast equipment issues weeks or even months before they occur. Farms no longer face unexpected downtime or lost productivity due to sudden machinery failures. Instead, maintenance can be scheduled during non-peak periods, ensuring minimal disruption to farm operations. This approach enhances efficiency and extends the lifespan of agricultural machinery.

Incorporating AI into equipment maintenance isn't just about preventing breakdowns. It also helps in optimizing the performance of the machinery. Precision tuning of equipment parameters based on AI recommendations ensures that machinery operates at its peak efficiency. Think of a combine harvester adjusted automatically for optimal grain quality and yield or irrigation systems modified on-the-fly to deliver precise water volumes. These advancements translate directly into cost savings and increased crop yields, catalyzing a more sustainable farming approach.

For tech-savvy farmers, the integration of AI systems is a game-changer. The data-driven insights that these systems provide are invaluable for decision-making. Advanced diagnostics coupled with predictive analytics enable farmers to prioritize repairs based on urgency and potential impact. This reduces the overall maintenance costs and ensures that machinery downtime is kept to a minimum. Scalability of these AI solutions is another point of interest. Whether a farmer manages a small plot or an extensive agricultural operation, AI models can be scaled up or down to meet the specific needs of various farming operations.

AI also plays a crucial role in the automation of routine maintenance tasks. Machine learning algorithms can identify repetitive and low-risk tasks that can be delegated to robotic systems. Automated lubrication, self-cleaning mechanisms, and real-time diagnostics are becoming increasingly common. This not only reduces the burden on farmers but also ensures consistency and precision in maintenance activities. Moreover, automation frees up valuable human resources, allowing farmers to focus on more strategic and high-value tasks, fostering innovation and enhancing productivity.

Notably, the integration of AI in equipment maintenance aligns with broader sustainability goals in agriculture. Optimally maintained equipment generally consumes less fuel and operates with greater efficiency, reducing the carbon footprint of farming activities. Additionally, AI-based systems can help in identifying underlying issues such as fuel leaks or inefficiencies in machinery operation, which might otherwise go unnoticed. Early detection and rectification of these issues contribute to more sustainable farming practices, promoting environmental conservation.

The agricultural equipment market is rapidly evolving to meet the demand for AI-enabled solutions. Leading manufacturers are incorporating AI technologies into new machinery, while retrofit kits are available for existing equipment. These advancements make it easier for farmers to adopt AI without needing to replace their entire fleet. The interoperability of AI systems with other technological solutions such as GPS-guided tractors and IoT devices further enhances their utility, creating a cohesive and integrated farming ecosystem.

Despite the undeniable advantages, the implementation of AI in equipment maintenance does come with challenges. The initial investment in AI technologies can be substantial, and there may be a learning curve associated with their adoption. However, as with any technological innovation, the long-term benefits often outweigh these initial hurdles. Additionally, continuous training and support from technology providers can help farmers maximize the benefits of AI and ensure seamless integration into their operations.

The reduction of downtime and increased machinery efficiency are perhaps the most immediate benefits farmers will notice with AI-driven maintenance systems. But there's also a less tangible yet significant advantage – peace of mind. Knowing that critical equipment is constantly monitored and that potential issues are flagged well in advance transforms how farmers approach their work. It's a shift from anxiety over inevitable breakdowns to a confident, proactive stance, enabling more strategic planning and execution of farm operations.

The intersection of AI and equipment maintenance is a prime example of how technology can revolutionize traditional practices in agriculture. By harnessing the power of AI, farmers can ensure their machinery is not only operational but optimized for maximum efficiency. This shift toward predictive maintenance is not just a technological upgrade; it's a step toward a more sustainable, resilient, and productive agricultural future.

In conclusion, AI's role in equipment maintenance and repair meaningfully contributes to the broader goals of precision agriculture and smart farming. The fusion of predictive analytics, automated diagnostics, and performance optimization forms the backbone of a futuristic yet practical approach to equipment management. As AI continues to evolve, its capabilities will only expand, offering even more sophisticated tools for farmers to maintain and improve their machinery. The future of farming, as it turns out, might depend not only on good soil and favorable weather but also on the algorithms humming within state-of-the-art agricultural equipment.

Reducing Downtime with Predictive Analytics

In the rapidly evolving landscape of modern agriculture, predictive analytics stands as a game-changing ally for minimizing equipment downtime. Utilizing the power of AI and machine learning, predictive maintenance strategies leverage data to foresee potential equipment failures before they transpire. This proactive approach ensures that machinery operates at peak efficiency while mitigating the risk of unexpected breakdowns that could disrupt critical farming activities.

Imagine a tractor plowing a field without hiccups, seamlessly transitioning from one task to the next. Predictive analytics makes this possible by continuously monitoring equipment health through data collected from various sensors embedded within the machinery. Vibration levels, temperature fluctuations, and operational performance metrics are just a few of the input data streams analyzed in real time. When anomalies are detected, an alert signals potential issues, allowing for maintenance to be scheduled at convenient times rather than during crucial farming periods.

This proactive maintenance is not just about dodging disruptions; it's also a significant cost-saving measure. By identifying and addressing issues before they become severe, farmers can avoid the steep expenses associated with major repairs and parts replacements. Consequently, the longevity of machinery is enhanced, enabling more extended periods of uninterrupted productivity. For instance, routine oil changes and component replacements conducted based on predictive insights can prevent engine failures and extend the life of the equipment.

Farmers are no strangers to the multitude of variables impacting their operations, from weather conditions to market demand. Adding equipment failures to this list is the last thing they need. Predictive analytics offers a layer of reliability that allows farmers to focus more on crop yields and less on equipment malfunctions. This level of dependability is a dream come true for anyone seeking to optimize their agricultural practices through technology.

Let's delve into the methodology. Predictive analytics employs complex algorithms to interpret historical and real-time data. By analyzing past failure patterns and current operational conditions, these algorithms predict when an issue is likely to occur. Machine learning models continuously update their understanding based on new data, improving their accuracy over time. This dynamic approach ensures that the insights generated are current and reliable, providing a robust framework for maintenance planning.

But how do these insights reach the farmer? Here lies the true brilliance of integrating AI into agriculture. Many predictive maintenance systems feature intuitive dashboards and mobile applications, presenting data in a user-friendly manner. Alerts and recommendations are often accompanied by actionable insights, guiding farmers on precise steps to take. This way, even those who may not be tech-savvy can easily understand and benefit from predictive analytics, making advanced maintenance strategies accessible to all.

The farming community is beginning to recognize the myriad advantages of predictive maintenance. Consider the experience of a large-scale corn farmer who integrated AI-driven predictive analytics into their operations. Prior to adoption, mechanical failures were a frequent and costly problem, often interrupting the planting season.

Post-adoption, the farmer reported a significant reduction in unplanned downtime and maintenance costs, leading to an increase in overall productivity and profitability. This case study is one among many demonstrating the tangible benefits of predictive maintenance in real-world agricultural settings.

Another fascinating aspect of predictive maintenance is its potential to foster more sustainable farming practices. Reduced machinery downtime translates to lower fuel consumption and less wear and tear on equipment. This not only saves costs but also minimizes environmental impact, contributing to a more sustainable agricultural model. By conserving resources and reducing waste, predictive maintenance supports the broader goal of eco-friendly farming.

Beyond individual farm equipment, predictive analytics can also be applied to fleets of machinery. For example, in large farming operations where multiple tractors, harvesters, and sprayers work concurrently, predictive maintenance can provide a centralized overview of the entire fleet's health. This holistic perspective enables coordinated maintenance schedules, optimizing the productivity of the entire operation. Machine operators can receive tailored alerts regarding their specific equipment, ensuring that each unit receives the attention it needs without unnecessary overlap or downtime.

Integrating predictive maintenance with other AI-driven agricultural technologies creates a cohesive ecosystem where every aspect of farming is optimized. When predictive analytics is combined with precision agriculture, soil health monitoring, and crop surveillance, the results are compelling. Each technology feeds data into the other, creating a comprehensive, interlinked system that can respond adeptly to the variables impacting farm operations.

A practical implementation of this synergy is seen in automated irrigation systems. By using predictive analytics, these systems can preemptively adjust watering schedules based on imminent equipment maintenance requirements, ensuring that crop hydration remains uninterrupted. This level of integration demonstrates the potential for predictive analytics to serve as a cornerstone in building smarter, more efficient agricultural systems.

While the benefits are clear, it's essential to acknowledge the challenges that come with implementing predictive analytics. Initial setup costs and the need for specialized knowledge to interpret and action predictive insights are common hurdles. However, the ROI on such investments tends to be favorable, making it a worthwhile endeavor for those committed to advancing their agricultural practices.

The agricultural sector is on the cusp of a technological transformation, largely driven by the innovative applications of AI. Predictive analytics is a prime example of how these advancements can directly benefit farm operations. Automation, efficiency, sustainability, and lower operational costs are just some of the substantial advantages that await those who embrace predictive maintenance. Looking forward, we can anticipate even more sophisticated applications as AI technologies continue to evolve.

For farmers, adopting predictive analytics is not merely a technological upgrade; it's a strategic decision that can significantly enhance the resilience and efficiency of their operations. As they steer through the complexities of modern farming, predictive maintenance stands as a vital ally, ensuring that their equipment remains reliable, their costs stay controlled, and their focus remains where it should be—on producing quality crops and driving sustainable agriculture forward.

Chapter 10:
Supply Chain Optimization

With AI at the helm, the supply chain in agriculture transforms from a complex labyrinth into a streamlined model of efficiency. By leveraging AI-driven analytics, real-time monitoring, and predictive algorithms, farmers and distributors can enhance logistics, ensuring that produce reaches markets at peak freshness. AI optimizes storage facilities by predicting spoilage and recommending the best conditions for each crop type, thereby minimizing waste. Moreover, intelligent routing systems can dynamically adjust transportation paths based on weather, traffic, and demand, reducing fuel consumption and ensuring timely deliveries. It's not just about getting products from point A to point B; it's about revolutionizing the entire journey to maximize profitability and sustainability. This fusion of technology and logistics marks not just an evolution but a revolution in how we perceive and handle agricultural supply chains.

Enhancing Logistics with AI

Supply chain optimization is essential for modern agriculture, as it ensures that food products move efficiently from farm to consumer, minimizing waste while maximizing profitability. By integrating artificial intelligence (AI) into logistics, the agricultural supply chain can experience unprecedented levels of efficiency and reliability. AI technologies offer tools that help manage everything from harvest scheduling to transportation routes, ensuring that perishable goods arrive fresh and on time. Let's delve into how AI is fundamentally transforming logistics in the agricultural supply chain.

One of the most significant contributions of AI in logistics is predictive analytics. Through machine learning algorithms, AI can forecast demand and supply trends accurately. These forecasts are based on a complex variety of data inputs, including weather conditions, market trends, historical sales data, and even social media activity. Farmers and distributors can use these insights to plan their operations more effectively, reducing the risk of overproduction or shortages. This predictive power leads to better inventory management and more strategic resource allocation.

AI also enhances the efficiency of transportation logistics. Advanced machine learning models can optimize routing for trucks and other delivery vehicles, taking into account real-time traffic data, road conditions, and fuel efficiency. This optimization means that products can be delivered faster and more reliably, reducing fuel consumption and greenhouse gas emissions. Companies can save money on transportation costs and contribute to environmental sustainability at the same time.

In addition to route optimization, AI-powered automation within logistics hubs and warehouses improves operational efficiency. Autonomous robots and drones can perform various tasks such as sorting, packing, and even delivering items within the warehouse. These machines work tirelessly and with precision, drastically reducing human error and operational bottlenecks. Automation not only speeds up the process but also ensures that goods are handled with the utmost care, preserving their quality.

AI can also play a crucial role in maintaining the cold chain, which is vital for the transport of perishable agricultural products. Smart sensors and IoT devices, backed by AI algorithms, can monitor the temperature and humidity levels in real-time, triggering alerts and adjustments as needed. This real-time monitoring helps in preventing spoilage, extending the shelf life of perishable goods, and ultimately reducing food waste. Ensuring that these products reach their destinations in optimal condition is particularly important for maintaining consumer trust and meeting health regulations.

Another application of AI in logistics is in the area of blockchain technology. Blockchain, a decentralized ledger system, can be greatly enhanced by AI to provide unparalleled transparency and traceability in the supply chain. This technology allows all parties involved, from farmers to retailers, to access real-time data concerning the status and history of a particular product. Consumers can verify the origin and quality of the goods they're purchasing by simply scanning a QR code. This added layer of verification enhances trust and promotes ethical farming practices.

AI's role in demand forecasting and inventory management cannot be overstated. Traditional methods of inventory management often rely on manual tracking and reactive responses to stock levels. With AI, these tasks become proactive and optimized. Algorithms can predict when stock will need replenishing, based on a multitude of variables such as seasonal trends, weather forecasts, and historical data. This level of precision helps in maintaining optimal inventory levels, reducing the risk of overstocking or stockouts, and ultimately saving costs.

Moreover, the integration of AI in logistics helps to streamline communication within the supply chain. In a typical agricultural supply chain, there are multiple stakeholders involved, from farmers to distributors to retailers. AI-driven platforms can facilitate seamless communication and coordination among these stakeholders, ensuring that everyone is on the same page. This improved communication helps in quicker decision-making, faster problem resolution, and overall smoother operations.

Risk management is another area where AI proves invaluable. The agricultural supply chain is fraught with risks, ranging from natural disasters to market volatility. AI systems can identify potential risks early by analyzing vast amounts of data from various sources. This proactive risk identification allows stakeholders to take preemptive actions to mitigate these risks. For example, if an AI system predicts a potential supply chain disruption due to weather conditions, contingency plans can be put in place well in advance to ensure that there is minimal impact on operations.

Nonetheless, the implementation of AI in logistics isn't without challenges. There are issues related to data privacy and security, especially when dealing with sensitive information across multiple stakeholders. However, many of these hurdles can be overcome with careful planning and implementation of robust cyber-security measures. Additionally, there is sometimes resistance to change from those accustomed to traditional methods. Educating stakeholders about the benefits and long-term gains of AI can help in overcoming this resistance.

Looking forward, the potential for AI to further revolutionize logistics in agriculture is immense. With continuous advances in AI technologies, including improvements in machine learning algorithms and better integration with other emerging technologies like blockchain and IoT, the future of agricultural logistics looks very promising. These technologies will not only make supply chains more efficient but also more resilient, adaptable, and sustainable.

Ultimately, the integration of AI in logistics within the agricultural supply chain is about creating a more efficient, transparent, and sustainable system. It's about ensuring that food gets from farm to table with minimal waste and maximum efficiency. By leveraging the capabilities of AI, farmers and supply chain managers can navigate the complexities of modern agriculture, meeting the demands of consumers while maintaining profitability. The journey of agricultural products from the fields to our plates will be smoother, swifter, and smarter, paving the way for a more sustainable and efficient agricultural ecosystem.

Improving Storage and Distribution

In the realm of supply chain optimization, improving storage and distribution is paramount. The intersection of artificial intelligence (AI) and agriculture has opened up new avenues that hold the potential for remarkable advancements in how agricultural products are stored and distributed. Optimizing these logistics not only ensures the freshness and quality of produce but also minimizes waste, reduces costs, and improves overall efficiency.

AI has introduced predictive analytics as a game-changing tool in this area. Predictive analytics leverage vast amounts of data to anticipate future outcomes, trends, and demands. For instance, farmers can now predict which crops will be in high demand months in advance, thanks to AI models analyzing historical weather patterns, market trends, and consumer behavior. This foresight allows for better planning in both planting and harvesting schedules, ultimately reducing excesses and shortages.

Temperature control is another critical aspect of storage that AI has revolutionized. Many perishables require specific temperature conditions to maintain their freshness and nutritional value. AI systems now enable real-time monitoring and control of storage environments. Sensors placed in storage facilities continuously collect data on temperature, humidity, and other relevant factors. AI algorithms analyze this data and make instant adjustments, ensuring optimal storage conditions. For instance, in cold storage facilities, AI can predict temperature fluctuations and preemptively make adjustments to avoid spoilage.

Moreover, AI-driven automated warehouses are becoming increasingly prevalent. These smart warehouses use robotics and AI to streamline operations. Robots can perform tasks such as sorting, packaging, and transporting goods more efficiently and with fewer errors than human workers. AI systems organize the layout of the warehouse dynamically, ensuring that frequently accessed items are more accessible and storage space is utilized optimally. This not only speeds up distribution but also reduces labor costs and human error.

Another significant innovation comes in the form of AI-driven inventory management. Traditional inventory systems often struggle with inaccuracies and inefficiencies. AI, on the other hand, uses real-time data to automate and optimize inventory levels. Machine learning algorithms can predict when stock levels for particular items are running low and trigger automatic reordering. They can also identify patterns of spoilage or overstock, allowing for more refined inventory control policies.

Transportation, a vital link in the distribution chain, has also seen substantial improvements through AI. Route optimization algorithms consider various factors, including traffic conditions, fuel consumption, delivery schedules, and even weather forecasts, to devise the most efficient routes. This reduces fuel costs, and delivery times, and minimizes the carbon footprint of transportation. Autonomous delivery vehicles and drones, powered by AI, are also on the horizon, promising to revolutionize how products are distributed, especially in remote or hard-to-reach areas.

Blockchain technology, when combined with AI, can enhance traceability and transparency in the supply chain. AI algorithms can process and analyze blockchain data to track every stage of the product lifecycle, from farm to table. This ensures that consumers receive products that are not only fresh but also ethically sourced and safe to consume. Any anomalies in the supply chain can be quickly identified and addressed, fostering trust and confidence among consumers.

Additionally, AI has a role in packaging and labeling, which are essential for both storage and distribution. Smart packaging, equipped with sensors and AI algorithms, can monitor the conditions of the produce inside and alert consumers and distributors if the product is nearing spoilage. Labels equipped with QR codes linked to AI-backed databases can provide detailed information about the product's origin, journey, and quality, enhancing transparency and consumer trust.

The application of AI in improving storage and distribution extends to resource management as well. Efficient use of resources like energy in storage facilities is crucial for cost savings and environmental sustainability. AI systems can optimize energy usage by predicting demand and adjusting energy consumption accordingly, reducing wastage and lowering operational costs. Likewise, water usage in storage facilities can be optimized using AI to ensure that it is used efficiently, further contributing to sustainable practices.

Collaboration between different stakeholders in the agricultural supply chain is also facilitated by AI. Data sharing among farmers, distributors, retailers, and consumers can create a more integrated and responsive supply chain. AI platforms can aggregate and analyze data from various sources, enabling more coordinated and informed decision-making. For instance, a retailer could communicate forecasted demand changes to farmers via an AI platform, allowing for adjustments in production to meet upcoming market needs.

The benefits of AI in storage and distribution go beyond operational efficiency. They extend to environmental and social impacts as well. By reducing waste and optimizing resource use, AI contributes to sustainability goals and lessens the agricultural carbon footprint. Furthermore, the enhanced efficiency can lower food prices, making healthy, fresh produce more accessible to a broader population, thus addressing food security issues.

Adapting to these technological advancements does require investment and training, but the long-term benefits dwarf the initial costs. Farmers, distributors, and retailers who embrace AI will find themselves at a competitive advantage, capable of responding swiftly to market demands and changes. Furthermore, as AI technology continues to evolve, its applications in storage and distribution will become even more sophisticated, paving the way for innovations we can scarcely imagine today.

In conclusion, the amalgamation of AI into storage and distribution processes is not just a luxury but a necessity for the modern agricultural supply chain. It brings a level of precision, efficiency, and sustainability that traditional methods simply cannot match. By harnessing the power of AI, we can ensure that agricultural products are handled and delivered in the most efficient and sustainable way possible, contributing to a more resilient and responsible food system.

Chapter 11:
Market Analysis and Decision Making

Harnessing the power of artificial intelligence for market analysis in agriculture opens up a realm of possibilities that's both exciting and transformative. By leveraging AI algorithms, farmers gain real-time insights into market trends, enabling them to make informed decisions about when to plant, harvest, and sell their crops. These data-driven decisions are crucial, reducing uncertainties and maximizing profitability. In an industry so dependent on unpredictable variables such as weather, pests, and fluctuating market demands, AI's capability to analyze vast amounts of data swiftly provides a competitive edge. Integrating market predictions with on-the-ground farming practices ensures a harmonious balance between production and demand, leading to sustainable and profitable agriculture. Ultimately, AI-driven market analysis empowers farmers to adapt proactively to ever-changing market conditions, ensuring resilience and sustainability in their operations.

AI in Market Predictions

The convergence of artificial intelligence (AI) and market predictions in agriculture promises a horizon teeming with possibilities. In an industry as unpredictable and variable as farming, even slight inaccuracies in crop demand, pricing, and market trends can lead to substantial financial losses. AI steps into the scene as a game-changer, infusing precision and predictive power into market analysis, ultimately guiding decision-making processes for farmers and investors alike.

Traditional methods of market predictions often rely on historical data and seasonal trends, which, while valuable, may not always be accurate given the rapidly changing climate and consumer behaviors. AI algorithms, on the other hand, are designed to analyze expansive datasets, incorporating variables from weather patterns to global economic conditions, and offer more nuanced and accurate forecasts. These algorithms continuously learn and adapt, enhancing their prediction capabilities over time.

Consider a farmer deciding what crop to plant for the upcoming season. While intuition and experience play crucial roles, AI can augment these with data-driven insights. For instance, AI systems can process information on previous crop yields, soil health, anticipated weather conditions, and market demand to recommend the most profitable crop to plant. By leveraging such insights, farmers can make informed choices, reducing risks and optimizing returns.

A key advantage of AI in market predictions is its ability to analyze market sentiment through social media, news articles, and trade reports. Natural language processing (NLP) algorithms can parse through vast amounts of unstructured data to gauge public opinion and potential demand shifts. For example, if there's an increasing trend in consumer preference for organic produce, AI can highlight this early, giving farmers a strategic edge to adapt their crop planning accordingly.

Beyond planting decisions, AI's market predictions extend to supply chain management. By forecasting demand more accurately, AI-driven systems can optimize inventory levels, reducing both shortages and surpluses. Real-time insights into market conditions enable better coordination between producers, distributors, and retailers. This synchronization not only leads to cost savings but also minimizes waste—a critical factor in sustainable agriculture.

The predictive prowess of AI is not limited to local markets. Global market predictions are equally vital, especially for farmers who export their produce. AI algorithms can analyze international trade policies, currency fluctuations, and geopolitical events to forecast global market conditions. By understanding these dynamics, farmers can better navigate the complexities of international trade, ensuring their produce reaches the best possible markets at the most opportune times.

Furthermore, AI-driven market predictions support price optimization strategies. By examining historical pricing data and real-time market trends, AI can recommend optimal pricing models that maximize profits while remaining competitive. This is particularly beneficial in volatile markets where prices can fluctuate dramatically due to unforeseen events such as natural disasters or economic downturns.

AI's ability to predict market disruptions is another invaluable asset. By monitoring a wide range of indicators, from weather anomalies to political unrest, AI can provide early warnings of potential disruptions. This allows farmers and supply chain managers to implement contingency plans and mitigate risks effectively. Such proactive measures are essential for maintaining stability in the agricultural sector.

Diversification is another strategy supported by AI in market predictions. By identifying emerging trends and market opportunities, AI can suggest diversification options that align with the farmer's capabilities and resources. Whether it's exploring alternative crops, value-added products, or new markets, AI's insights empower farmers to innovate and expand their business horizons.

AI in market predictions also plays a pivotal role in financial planning. Accurate predictions enable farmers to make informed decisions about investments, loans, and insurance. Financial institutions can also use AI-driven market insights to assess risks and offer tailored financial products to farmers, fostering a more resilient and financially stable agricultural sector.

A challenge to consider is the accessibility of AI technologies to all farmers, especially smallholder farmers who may lack the resources to invest in sophisticated AI systems. However, collaborative efforts between governments, NGOs, and private sector companies can bridge this gap. By providing affordable access to AI tools and training, these stakeholders can democratize the benefits of AI in market predictions, ensuring even the smallest farms can thrive in a data-driven agricultural landscape.

The journey of integrating AI into market predictions is ongoing and ever-evolving. The potential benefits are immense, but it requires continuous adaptation and innovation. As AI technologies advance, their predictive accuracy and utility in agricultural market predictions will only grow, offering farmers an indispensable tool for navigating the complexities of modern agriculture.

The impact of AI in market predictions transcends beyond individual farms to influence the entire agricultural ecosystem. By enhancing market efficiency, reducing waste, and promoting sustainable practices, AI contributes to a more resilient and robust agricultural industry. This, in turn, supports global food security and the well-being of communities dependent on farming.

In conclusion, embracing AI in market predictions is not merely an option but a necessity for the future of agriculture. By harnessing the power of AI, farmers can transform uncertainties into opportunities, fostering a prosperous and sustainable agricultural sector for generations to come. The transformative potential of AI holds the promise of driving the next green revolution, ensuring that the global agricultural landscape is prepared to meet the challenges and opportunities of the future.

Data-driven Decision Making

In the realm of modern agriculture, the importance of data-driven decision making cannot be overstated. This transformative approach leverages the vast amounts of data generated by numerous technological advancements to guide farming practices, increase efficiency, and maximize yields. The essence of data-driven decision making lies in its ability to sift through enormous datasets to extract actionable insights. This is where artificial intelligence (AI) and machine learning (ML) come into play, revolutionizing how farmers understand and interact with their land and crops.

Historically, agricultural decisions were often driven by intuition, experience, and tradition. While these elements still hold value, they are no longer sufficient in the face of contemporary challenges such as climate change, resource scarcity, and the need for sustainable practices. Precision agriculture, a concept explored earlier, feeds data directly into decision-making processes, enhancing the accuracy and profitability of agricultural operations. With AI, data is not merely collected but also analyzed in real-time, allowing for more immediate and precise interventions.

One vivid example of data-driven decision making can be seen in crop health monitoring. By integrating AI systems that analyze weather patterns, soil conditions, and aerial imagery, farmers can predict disease outbreaks before they become widespread. AI algorithms can process the complex interplay between various environmental factors, highlighting risks and suggesting preventative measures. This proactive approach not only saves crops but also reduces the reliance on pesticides, contributing to more sustainable farming practices.

The power of big data in agriculture extends beyond individual farms. Aggregated data from multiple sources, including satellite imagery, weather forecasts, and market trends, can inform regional and even global agricultural strategies. For instance, predictive models can anticipate market demand for certain crops, allowing farmers to adjust their planting schedules accordingly. This level of foresight ensures that supply meets demand, minimizing waste and maximizing profit.

The integration of AI in soil analysis has also revolutionized farming. Traditional soil testing methods are often labor-intensive and time-consuming. In contrast, AI-driven soil testing can quickly and accurately measure nutrient levels, pH balance, and moisture content. Predictive analytics can then determine the optimal times for planting, watering, and fertilizing. Such precision minimizes input costs and environmental impact while ensuring healthier, more robust crops.

Water management is another critical area where data-driven decision making is making significant strides. AI systems can predict rainfall patterns and soil moisture levels with remarkable accuracy. These predictions allow farmers to optimize irrigation schedules, ensuring that crops receive just the right amount of water at the right time. This not only conserves water but also prevents issues like soil erosion and nutrient leaching.

The integration of data-driven decision making in supply chain logistics exemplifies the broad applicability of AI in agriculture. By analyzing data on crop yields, transportation routes, and market prices, AI can streamline the distribution process. Farmers can determine the most efficient pathways for their produce, reducing transportation costs and ensuring fresher products reach consumers quickly. In turn, this enhances profitability and sustainability across the entire supply chain.

The beauty of data-driven decision making lies in its continuous improvement cycle. As more data is collected, the algorithms become more sophisticated and accurate. For instance, machine learning models can be trained on vast datasets to improve their predictive capabilities over time. This iterative process ensures that the system adapts to changing conditions and continually provides the most relevant insights.

The transition to data-driven decision making does come with challenges. One major obstacle is the digital divide, particularly in rural and underdeveloped areas where access to technology and internet connectivity may be limited. Bridging this gap is essential to ensure that all farmers, regardless of their location, can benefit from AI-driven insights. Collaborative efforts between governments, NGOs, and tech companies are crucial in providing the necessary infrastructure and training to support this technological shift.

Beyond logistical challenges, there is also the need for education and training. Farmers must become adept at interpreting and acting on the data provided by AI systems. This requires a shift in mindset from traditional practices to a more analytical approach. Training programs and workshops focused on data literacy and AI applications in agriculture can empower farmers to make informed decisions confidently.

Data-driven decision making doesn't just transform individual farms; it has the potential to reshape entire agricultural ecosystems. By enabling more precise and responsive farming practices, it can lead to better resource management, reduced environmental impact, and increased food security. In regions affected by climate change, data-driven insights can guide adaptive strategies, ensuring resilience in the face of unpredictable conditions.

Moreover, the ethical considerations surrounding data usage cannot be ignored. Farmers must have ownership and control over their data to prevent exploitation and ensure privacy. Policies and regulations need to be in place to safeguard against misuse and to promote transparency and fairness in data management practices.

The future of agriculture lies in the seamless integration of AI and data-driven decision making. As we move forward, the possibilities are endless. Precision in farming will continue to improve, resource usage will become more efficient, and the agricultural sector will be better equipped to meet the demands of a growing population. By embracing this technology, farmers can transform their practices, leading to a more sustainable, productive, and prosperous agricultural future.

In conclusion, data-driven decision making is a pivotal component in the evolution of modern agriculture. It represents a shift from intuition-based farming to a systematic, empirical, and highly efficient approach. By harnessing the power of AI and big data, farmers can navigate the complexities of today's agricultural challenges, unlocking new levels of productivity and sustainability. This transformative potential is the harbinger of the next green revolution, paving the way for a thriving, resilient, and food-secure world.

Chapter 12:
Smart Greenhouses

Smart greenhouses represent a transformative leap in agricultural technology, marrying traditional horticulture with cutting-edge AI innovations. By integrating sophisticated machine learning algorithms, sensor networks, and automated systems, these controlled environments ensure optimal growth conditions for crops year-round. Imagine a space where temperature, humidity, light, and nutrient supply are continuously monitored and perfectly adjusted in real-time—this is the promise of smart greenhouses. They can predict and mitigate issues like pest infestations or nutrient deficiencies before they become problematic. Beyond just enhancing productivity, smart greenhouses reduce resource consumption, making farming more sustainable and profitable. The implications for food security are profound, enabling higher yields and more resilient crop production even amidst climate variabilities. Indeed, the global shift toward intelligent, self-regulating greenhouses could be the linchpin in ushering in the next green revolution, making high-tech, sustainable farming accessible to producers of all sizes.

Integrating AI in Controlled Environments

Smart greenhouses are transforming agriculture by merging technology with traditional farming practices. Controlled environments already offer significant advantages, such as protection from the elements, optimized growing conditions, and the ability to produce crops year-round. When artificial intelligence (AI) is integrated into these controlled environments, the results are groundbreaking. In essence, AI has the capability to create hyper-efficient agricultural systems that can address some of the most pressing challenges in modern agriculture.

The core objective of integrating AI in controlled environments is to automate and optimize every aspect of greenhouse operations. This includes environmental control, crop management, and resource allocation. AI algorithms can analyze data from an array of sensors placed throughout the greenhouse, monitoring variables like temperature, humidity, light intensity, and CO_2 levels. The precision with which AI can manage these parameters ensures optimal growing conditions, leading to healthier plants and higher yields.

Consider temperature control. While traditional systems rely on periodic manual adjustments, AI can make real-time, incremental changes based on predictive analytics. For instance, if the system detects that sunlight levels will rise significantly in the next hour, it can preemptively adjust cooling mechanisms just enough to maintain a stable temperature. Such fine-tuned adjustments can result in significant energy savings and more consistent crop growth.

Light management in smart greenhouses has also been revolutionized by AI. AI-driven systems can use machine learning to adapt artificial lighting schedules to the needs of specific crops, accounting for variations in natural light availability. By optimizing the spectrum and intensity of artificial light, plants receive exactly what they need for photosynthesis, maximizing growth rates and improving overall plant health.

Water management is another critical area where AI shines. Traditional irrigation methods can be wasteful, often applying a uniform amount of water regardless of individual plant needs. AI-enabled systems, however, can assess the moisture levels of the soil in real-time and adjust water distribution accordingly. This not only conserves water but also ensures that plants receive adequate hydration, thus reducing stress and enhancing growth.

In addition to environmental controls, AI plays a pivotal role in crop monitoring within smart greenhouses. Advanced imaging technologies combined with AI algorithms allow for real-time surveillance of plant health. These systems can detect early signs of diseases, nutrient deficiencies, or pest infestations, prompting immediate intervention. By catching these issues early, farmers can mitigate damage and minimize crop loss, leading to more sustainable farming practices.

Nutrient management, often a tedious and imprecise task, becomes highly efficient when aided by AI. Through data collected from various sensors, AI can formulate precise nutrient recipes tailored to the specific growth stage and health condition of each crop. Constant adjustments ensure optimal nutrient uptake, avoiding deficiencies or toxicities and resulting in healthier, more productive plants.

AI-powered decision support systems also empower farmers to make more informed choices. These systems can analyze vast amounts of data, from historical crop performance to current market trends, and generate actionable insights. For example, AI can forecast the most profitable planting cycles or recommend crop varieties that are likely to thrive under specific controlled conditions. This blend of historical data with real-time analytics enables smarter planning and risk management.

One of the most forward-thinking applications of AI in smart greenhouses is the concept of predictive analytics. By analyzing patterns in data over time, AI can forecast potential issues before they arise. For instance, it might predict a sudden drop in humidity based on upcoming weather patterns and adjust the greenhouse's systems to mitigate any potential negative impact on crops. This level of foresight is invaluable in preventing disruptions and maintaining a stable growth environment.

Furthermore, integrating AI with autonomous machinery elevates the operational efficiency of smart greenhouses. Drones and robots, equipped with AI algorithms, can handle tasks such as planting, harvesting, and maintenance with precision. These machines can work around the clock, performing repetitive tasks with consistent accuracy, which not only enhances productivity but also frees up human labor for more complex decision-making tasks.

Artificial intelligence also fosters sustainability in controlled environments. Energy management systems driven by AI can reduce greenhouse energy consumption by optimizing the use of lights, fans, and heaters. Additionally, by precisely controlling water and nutrient delivery, AI minimizes waste, further promoting sustainable practices. These efficiencies are crucial as the agricultural sector strives to meet global food demand while mitigating environmental impact.

Collaborative AI platforms bring yet another dimension to the realm of smart greenhouses. These platforms allow multiple greenhouses to share data and insights, creating a collective intelligence that benefits all participants. For example, if one greenhouse encounters a particular pest infestation and develops an effective countermeasure, this information can be shared instantaneously with connected greenhouses, enabling a rapid, coordinated response.

Despite the myriad benefits, integrating AI in controlled environments doesn't come without challenges. Initial setup costs can be significant, and there is a learning curve associated with adopting new technologies. However, the long-term benefits—such as increased yields, resource efficiency, and reduced labor costs—often outweigh these initial hurdles. Moreover, ongoing advancements in AI technology and declining costs of hardware components are making these systems more accessible to a wider range of farmers.

The integration of AI in controlled environments isn't just about technological advancement; it's about rethinking the way we grow our food. By leveraging AI, we can move towards a more efficient, sustainable, and resilient agricultural system. This is particularly significant in the face of global challenges like climate change, population growth, and resource scarcity. Smart greenhouses, powered by AI, represent a beacon of hope for a future where technology and nature coexist harmoniously to feed the world.

Looking ahead, the potential of AI in smart greenhouses is boundless. Future innovations could lead to fully autonomous greenhouses, where AI not only optimizes existing processes but also evolves and learns from its environment. These self-improving systems could push the boundaries of agricultural productivity, sustainability, and environmental stewardship to unprecedented levels.

In conclusion, AI's integration into controlled environments like smart greenhouses is a transformative force in modern agriculture. By providing precise control over environmental conditions, enhancing crop monitoring, optimizing resource use, and supporting informed decision-making, AI helps create more efficient, sustainable, and productive agricultural systems. As we continue to explore and innovate, the synergy between AI and controlled environments promises to be a cornerstone of the next agricultural revolution.

Optimizing Growth Conditions

Smart greenhouses represent a revolution in agricultural technology, blending artificial intelligence (AI) and advanced engineering to create ideal growth environments for crops. The essence of optimizing growth conditions lies in harnessing technologies to monitor, adjust, and perfect the environments inside these greenhouses. This effort is multifaceted, encompassing temperature control, humidity regulation, light management, and nutrient delivery. Each element plays a pivotal role in maximizing plant health and productivity.

Temperature control in smart greenhouses is a cornerstone of optimizing growth conditions. Traditional greenhouses rely heavily on manual adjustments and basic thermostats. However, smart greenhouses deploy AI-driven systems equipped with sensors to continuously monitor and modulate temperatures. These systems analyze data in real-time and make micro-adjustments to maintain optimal conditions for plant growth. For example, if the internal temperature begins to veer outside the ideal range, the system can automatically activate cooling or heating mechanisms, ensuring that the crops never experience thermal stress.

Humidity is another critical factor in plant health that smart greenhouses manage with precision. Maintaining the right humidity levels can prevent diseases and enhance plant growth. AI systems come into play here by analyzing real-time data from humidity sensors and automatically adjusting misting systems, vents, or dehumidifiers. This ensures that the microclimate within the greenhouse remains stable, creating an environment where plants can thrive without the threat of mold, mildew, or other diseases that thrive in poor humidity conditions.

The integration of AI has also transformed how light is managed within smart greenhouses. Plants have specific light requirements that vary by species and growth stage. Advanced LED lighting systems can simulate natural sunlight, but with the added benefit of controllability. AI algorithms can adjust the spectrum and intensity of light according to the plants' needs, time of the day, and weather conditions. This fine-tuning not only boosts photosynthesis but also optimizes energy efficiency. Photoperiodism—the response of plants to the length of day—is key to certain crops, and AI ensures that this is meticulously managed for improved yields.

Automated nutrient delivery systems bring another level of precision to smart greenhouses. Traditional fertilization methods can be wasteful and inconsistent. In contrast, AI-driven systems can tailor nutrient supply based on real-time soil and plant data. Sensors measure various parameters like pH, nutrient concentration, and moisture levels, feeding this data back to a central AI system. It calculates the exact nutrient mix needed and delivers it through automated irrigation systems, ensuring plants receive the right nutrients at the right time. This reduces waste and minimizes the environmental footprint of fertilization.

Furthermore, these intelligent systems are capable of learning and adapting. Machine learning algorithms analyze years of collected data to continually improve their predictions and actions. This means the greenhouse environment becomes progressively more optimized over time, as the system learns what works best for specific plants under various conditions. By integrating historical data with real-time analytics, AI can forecast potential issues and preemptively adjust growing conditions to mitigate them.

Maintaining a delicate balance of conditions is particularly crucial when considering the variety of crops that can be grown in smart greenhouses. Different plants have unique requirements, and a one-size-fits-all approach is inefficient. AI allows multiple zones within the same greenhouse to operate under different conditions tailored to specific crops. This zonal optimization enables growers to cultivate a diverse array of plants simultaneously, maximizing the use of space and resources.

Smart greenhouses also leverage predictive analytics to anticipate and prepare for external environmental changes. By accessing weather forecast data, these greenhouses can make proactive adjustments to internal conditions. For instance, if a cold spell is predicted, the system can preheat the greenhouse or adjust humidity levels to counteract the expected drop in temperature. This foresight minimizes risks associated with sudden environmental changes and shields the crops from potentially harmful conditions.

Integrating AI for optimizing growth conditions also facilitates enhanced resource management. Efficient use of water, energy, and nutrients contributes to both economic and environmental sustainability. Through precise control and reduction of wastage, smart greenhouses can operate with lower input costs while yielding higher outputs. This efficiency translates to cost savings for farmers and reduced environmental impact—essential for the future of sustainable agriculture.

Real-time monitoring and control also improve the reliability and quality of produce. By ensuring crops grow in ideal conditions, smart greenhouses can consistently produce high-quality yields that meet market standards. This reliability is crucial for farmers to build trust with buyers and sustain profitable operations.

Moreover, the data generated within smart greenhouses is invaluable for research and development. Scientists and agronomists can study this data to understand plant behavior better, leading to the development of improved crop varieties and better farming practices. This continuous cycle of data collection, analysis, and application fosters innovation in agricultural practices.

Optimizing growth conditions in smart greenhouses extends beyond individual pods of crops; it scales up to large commercial operations. Scalable AI solutions make it feasible to manage vast agricultural enterprises with the same level of precision and care as smaller operations. This scalability is vital for meeting the increasing global food demand while reducing the agricultural footprint.

The social implications of optimized growth conditions are equally significant. By enabling the production of more food with fewer resources, smart greenhouses can play a crucial role in combating food insecurity. They can be deployed in urban areas, reducing the need for food transportation and providing fresh produce locally. This urban agriculture not only reduces traffic and emissions but also supports local economies and fosters community engagement.

The journey of optimizing growth conditions in smart greenhouses is both an ongoing challenge and a testament to human ingenuity. By continually refining AI technologies and integrating them into agricultural practices, we move closer to a future where farming is more efficient, sustainable, and capable of feeding the growing global population. The innovative spirit driving these advancements serves as an inspiration, reminding us that the marriage of technology and nature can yield extraordinary results.

Chapter 13:
Livestock Management

The integration of artificial intelligence in livestock management has revolutionized traditional farming practices, significantly enhancing efficiency and productivity. By leveraging AI-driven technologies, farmers can now track animal health and productivity with unprecedented accuracy through advanced monitoring systems that detect early signs of illness, nutritional deficiencies, and behavioral changes. This proactive approach not only optimizes animal welfare but also boosts yield and quality. Automated feeding and milking systems have further streamlined operations, ensuring precise nutrient delivery and milking schedules tailored to each animal's needs. Imagine a farm where every cow's daily routine is meticulously managed by smart algorithms, freeing up time for farmers to focus on strategic planning and innovation. This synergy of technology and agriculture paves the way for a more sustainable and profitable future, highlighting the immense potential of AI to drive the next green revolution in livestock management.

AI in Tracking Animal Health and Productivity

The application of artificial intelligence in tracking animal health and productivity represents a seismic shift in livestock management. By leveraging AI technologies, farmers can now monitor various health parameters and productivity metrics with unprecedented precision, enabling proactive interventions and optimizing animal welfare. This evolution is transforming traditional farming into a more sustainable, efficient, and responsive industry.

AI tools, including machine learning algorithms and IoT sensors, facilitate real-time monitoring of livestock. These technologies gather data on various health indicators such as temperature, heart rate, feeding patterns, and movement. Smart collars and wearable devices transmit this information to centralized systems, where AI algorithms analyze the data, flagging anomalies that might indicate illness or distress. This continuous health surveillance helps in early disease detection, often before symptoms become visible to the naked eye.

In addition to health monitoring, AI plays a crucial role in tracking productivity metrics like milk yield, weight gain, and reproductive performance. For dairy farmers, AI-powered sensors can measure milk quality and quantity during each milking session, providing immediate feedback on the health and productivity of individual cows. Such granular insights help farmers understand the link between health and productivity, driving better management decisions.

Advanced machine learning models can predict and optimize growth rates by analyzing historical and real-time data. By feeding data into predictive models, farmers can forecast growth trajectories and adjust feeding regimes, ensuring livestock reach their full potential. Predictive analytics also help in planning breeding cycles, enhancing genetic quality, and improving overall herd productivity.

One notable example is the use of AI-powered cameras and sensors to conduct behavior analysis. Changes in an animal's behavior, such as reduced mobility or altered feeding patterns, can be early indicators of health issues. AI algorithms process images and videos from these devices to detect such changes, offering a non-invasive means to monitor animal welfare continually.

Furthermore, AI's ability to integrate various data sources enhances the accuracy of health and productivity tracking. Environmental factors, such as weather conditions and barn microclimates, can be correlated with health data to identify stressors impacting animal welfare. For example, heat stress in summer months can significantly reduce milk production in dairy cows. AI systems can predict heatwaves and recommend preemptive measures like cooling systems or adjusted feeding schedules to mitigate the impact.

In modern farming, the combination of AI and genomic data is creating new frontiers. Genomic selection, supported by AI, allows farmers to identify animals with desirable traits more accurately, thus driving genetic improvements. By analyzing genetic markers and correlating them with performance data, AI can help farmers select animals with higher disease resistance, better growth rates, and improved fertility, thereby accelerating the breeding of superior livestock.

Another transformative application is in automated health scoring systems. These AI-driven systems score the health of livestock based on various parameters, providing a comprehensive view of an animal's condition. Health scoring is particularly valuable in large-scale operations where manual health checks are impractical. Automated systems ensure that all animals are regularly monitored, and those requiring attention are identified promptly.

Farm management software integrated with AI capabilities is also playing a pivotal role. These platforms aggregate data from various sources, providing a unified dashboard that offers actionable insights. Smart algorithms analyze trends, predict outcomes, and suggest interventions, facilitating data-driven decision-making. For example, if a particular health issue is detected in a subset of the herd, the system can identify common factors and recommend preventive measures for the rest of the animals.

The benefits of AI in livestock management extend beyond individual farms. At the industry level, aggregated data from numerous farms can reveal patterns, contributing to broader disease control and productivity enhancement strategies. Regional and global monitoring can aid in early detection of outbreaks and help in formulating effective responses, thereby safeguarding public health and animal welfare on a larger scale.

Despite these advancements, adopting AI in livestock management does present challenges. High initial costs and the need for technical expertise can be barriers, particularly for smaller farms. Additionally, the quality and consistency of data play a critical role in the efficacy of AI systems. Ensuring accurate data collection, minimizing noise, and addressing data gaps are essential for reliable AI-driven insights.

Ethical considerations also come into play. Data privacy and the welfare of animals must be addressed to ensure that the deployment of AI technologies is responsible and aligned with societal values. Transparent practices and adherence to ethical guidelines can help in gaining the trust of farmers and the broader community, facilitating wider adoption of AI solutions.

Looking ahead, advancements in AI, combined with emerging technologies like blockchain, promise even greater transformative potential. Blockchain can enhance traceability and authenticity in livestock management, ensuring that data used for AI analytics is genuine and tamper-proof. This synergy of AI and blockchain can create robust systems for tracking animal health and productivity, fostering trust and accountability in the agri-food supply chain.

In conclusion, AI is revolutionizing how we understand and manage livestock health and productivity. By providing deeper insights and facilitating proactive care, it is helping farmers optimize their operations and improve animal welfare. As these technologies continue to evolve, we can expect even more innovations that will drive the next phase of growth in sustainable and efficient livestock management.

Automated Feeding and Milking Systems

In the rapidly evolving sector of livestock management, automated feeding and milking systems have emerged as game-changers. The integration of AI and robotics into these systems isn't merely an enhancement—it's a revolution. These innovations bring about unprecedented efficiency and precision, addressing both the economic and labor challenges traditionally faced by the farming industry.

The role of automated feeding systems is paramount. Leveraging AI, these systems ensure that each animal receives the precise amount and type of feed needed for optimal growth and milk production. Traditionally, feeding livestock has been a labor-intensive and imprecise task. Farmers had to rely on manual methods, which often resulted in either overfeeding or underfeeding, leading to wasted resources and suboptimal animal health. Automated systems circumvent these issues by using advanced sensors and algorithms to tailor feeding schedules to the specific needs of each animal.

One of the most significant benefits of these automated feeding systems is the precision in nutrient delivery. Animal growth and milk production are inherently influenced by the quality and quantity of feed they consume. By employing AI-driven data analysis, modern feeding systems can calculate the nutritional needs based on an array of variables such as age, weight, production stage, and health status. The result is a finely tuned feeding regimen that maximizes productivity while minimizing waste.

Milking systems, like their feeding counterparts, have undergone a dramatic transformation through AI integration. Automated milking systems (AMS) or robotic milking systems (RMS) were designed with both animal welfare and farmer convenience in mind. These systems can operate around the clock, freeing up farmers from the rigid schedules demanded by traditional milking practices. The constant availability of milking robots encourages cows to be milked multiple times a day, as they feel comfortable, rather than sticking to a strict schedule imposed by human intervention. This natural rhythm can lead to increased milk yield and improved overall herd health.

AI plays a crucial role in the functioning of these AMS. For instance, machine learning algorithms analyze data from each milking session—such as the amount of milk produced, the time taken for milking, and any signs of distress or illness in the cows. This data isn't just collected; it's processed in real-time to provide actionable insights. Deviations from normal patterns can signal potential health issues, allowing for early intervention before a minor problem escalates into something more severe.

A major advantage of RMS is their ability to integrate with other farm management systems. The interoperability of these technologies means that data collected from feeding, milking, and health monitoring systems can be aggregated and analyzed holistically. This 360-degree view of livestock operations helps farmers make informed decisions, optimizing not just individual animal health, but the overall farm productivity. For instance, adjusting feeding protocols based on milk production data ensures that nutrition is linked directly to output, driving both efficiency and profitability.

Yet, it's not just large-scale dairy farms that can benefit from these innovations. Smaller farming operations, often with tighter budgets and fewer hands, stand to gain significantly as well. The scalability and cost-effectiveness of modern AI systems make them accessible even to those with limited resources. This democratization of technology helps level the playing field, providing smaller farmers with tools that were once the domain of large, industrial farming operations.

Addressing the environmental impact of livestock farming remains a critical concern. Automated systems contribute positively in this area by reducing feed waste and optimizing resource usage. Precise feeding systems ensure that every ounce of feed is utilized efficiently, which translates to less waste and lower environmental footprint. Moreover, healthier, well-fed animals tend to produce more milk and have longer productive lives, reducing the need for frequent herd replacements and resource-intensive farming practices.

Looking at animal welfare, automated feeding and milking systems offer considerable improvements. Animals benefit from consistent, stress-free routines. Automated systems reduce human error and the physical toll of manual labor, fostering a more humane treatment of livestock. The AI-driven data analytics assist in early disease detection and herd health management, ensuring timely medical intervention and reducing suffering.

One might wonder about the economic considerations of implementing such advanced technologies. While the initial investment in AI-driven automated systems might seem steep, the long-term benefits notably outweigh the costs. Enhanced productivity, reduced labor costs, minimized waste, and improved animal health collectively contribute to a better return on investment. Many farming operations report significant improvements in profitability within just a few seasons of adopting these technologies.

In conclusion, the integration of AI into automated feeding and milking systems embodies the future of livestock management. These technologies, through precision, efficiency, and data-driven insights, are transforming the way farmers manage their herds. They are not merely tools for convenience; they represent a paradigm shift towards more sustainable, profitable, and humane farming practices. As these systems continue to evolve, fostering a deeper synergy between technology and agriculture, the potential benefits for both the industry and the planet are limitless.

Chapter 14:
Sustainability and Environmental Impact

Adopting AI in agriculture isn't just a revolution in efficiency and productivity; it's a significant stride toward environmental sustainability. Picture farms where AI optimizes water usage, significantly reducing wastage, and sensors monitor soil health to minimize the need for chemical fertilizers. Advanced algorithms predict and manage pest invasions, ensuring that interventions are both effective and ecologically sound. By smartly aligning agricultural practices with nature's rhythms, AI helps reduce carbon footprints and promote biodiversity. It's about weaving technology and farming into a harmonious fabric that not only feeds the world but also nurtures the planet, ensuring the prosperity of future generations. Through AI's environmental impact, we find a sustainable path that reshapes agriculture from a stressor to a steward of our ecosystems.

Reducing Carbon Footprint with AI

Cutting down on carbon emissions is not just a goal, it's a necessity, especially in agriculture. Traditional farming practices have long been contributors to greenhouse gases through methods such as tillage, excessive fertilization, and uncontrolled livestock farming. However, the advent of artificial intelligence (AI) offers a host of innovative solutions to reduce the carbon footprint of agricultural practices.

One of the most promising areas where AI can make a profound impact is in precision agriculture. By optimizing the use of resources like water, fertilizers, and pesticides, AI-driven systems minimize waste and reduce the carbon emissions associated with their production and application. For instance, AI algorithms can analyze extensive datasets to determine the exact amount of fertilizer a particular plot of land needs, thereby reducing the volumes that would otherwise contribute to greenhouse gas emissions.

Moreover, AI can assist in carbon sequestration efforts, which are vital for capturing atmospheric carbon and storing it in soil and plants. Advanced machine learning models can analyze soil composition data and recommend practices that enhance the soil's ability to retain carbon. Techniques such as reduced tillage and cover cropping can be optimized using AI to maximize their carbon-sequestering potential.

Renewable energy integration in agricultural practices is another key avenue for reducing carbon emissions, and AI plays a pivotal role here too. AI can optimize the operation of renewable energy systems like solar panels and wind turbines to ensure they function at maximum efficiency. This not only reduces reliance on fossil fuels but also ensures that the energy consumption patterns are adjusted dynamically to minimize waste and maximize utility.

AI-powered crop management systems also contribute to reducing the carbon footprint by predicting and preventing crop diseases. Early detection systems powered by AI can identify disease outbreaks before they become widespread, significantly reducing the need for chemical treatments. This decreases the production and application of chemicals, which are both energy-intensive processes, thereby lowering greenhouse gas emissions.

Another significant contribution of AI is in optimizing logistics within the agricultural supply chain. AI-driven logistics solutions can streamline transportation routes, reduce fuel consumption, and improve overall efficiency. When AI algorithms predict the best routes and methods for transporting agricultural products, it reduces the carbon footprint associated with getting goods to market.

Autonomous machinery powered by AI also provides a more sustainable alternative to traditional, human-operated machinery. These machines are more efficient and can work with a level of precision that reduces fuel consumption and minimizes soil compaction, leading to healthier, carbon-retaining soils. Autonomous tractors, for instance, can be programmed to optimize their routes to use the least amount of fuel, thereby emitting fewer greenhouse gases.

Even at the level of irrigation, AI offers transformative potential. Traditional irrigation methods often waste water and the energy used to pump it. AI systems can analyze soil moisture and weather data in real-time to provide just the right amount of water needed for crops, thus conserving water and reducing the energy spent on irrigation. This not only lowers the carbon footprint but also enhances the sustainability of water resources.

Reducing livestock emissions is another critical aspect where AI can make a difference. Advanced AI algorithms can predict the nutritional needs of livestock more accurately, thereby optimizing feed efficiency and reducing methane emissions from animals. Furthermore, AI can monitor the health of livestock using sensors and cameras, enabling early interventions that can prevent diseases and reduce the need for antibiotics and other treatments that contribute to greenhouse gas emissions.

The concept of smart greenhouses is yet another frontier where AI's impact on reducing carbon footprint cannot be overstated. AI systems control the environment inside greenhouses, ensuring optimal conditions for plant growth. By precisely regulating temperature, humidity, and CO_2 levels, AI not only enhances productivity but also ensures that resources are used as efficiently as possible, thereby reducing overall carbon emissions.

Furthermore, predictive maintenance enabled by AI helps in managing agricultural equipment in a way that minimizes emissions. By predicting equipment failures before they occur, AI systems ensure that machinery operates at optimal efficiency levels, reducing unnecessary repairs and replacements that add to the carbon footprint. Predictive maintenance thus extends the longevity and efficiency of farm equipment.

The integration of AI in sustainable agriculture practices also holds promise for broader climate adaptation strategies. AI can analyze climatic and environmental data to predict future trends, allowing farmers to adapt their practices in ways that lower emissions. For example, AI can assist in choosing crop varieties that are more resilient to changing climatic conditions, thereby reducing the need for resource-intensive interventions.

Finally, the role of AI in enhancing market analysis and decision making ensures that agricultural practices align with sustainable goals. AI algorithms can predict market demands more accurately, allowing farmers to plan their crop cycles efficiently and reduce overproduction, thereby minimizing waste and associated carbon emissions. Data-driven decision-making facilitated by AI ensures that every aspect of agricultural operations is optimized for sustainability.

In summary, AI offers a multifaceted approach to reducing the carbon footprint of agriculture. From optimizing resource usage and enhancing carbon sequestration to improving efficiency in logistics and machinery operations, AI technologies present a compelling solution to one of the most pressing challenges of our time. By integrating AI into various agricultural processes, we not only pave the way for a more sustainable future but also contribute significantly to the global efforts in combating climate change.

Promoting Biodiversity and Conservation

In the contemporary agricultural landscape, promoting biodiversity and conservation isn't just an ethical priority; it's a strategic necessity. The monocultures that once dominated agriculture have led to a plethora of environmental issues, from soil degradation to vulnerability to pests and diseases. Modern farming practices, embedded with artificial intelligence (AI), offer the promise of reversing this trend. By leveraging sophisticated algorithms and real-time data, AI can help create a more biodiverse and resilient ecosystem.

AI's role in promoting biodiversity begins with its capacity to analyze and interpret vast datasets. For instance, machine learning algorithms can examine historical data on crop yields, soil health, and weather conditions to suggest diversifying crop rotations. This isn't just about planting different crops in different seasons; it's about creating a mosaic of plant species that can mutually benefit each other. Diverse cropping systems can enhance soil fertility, reduce pest outbreaks, and improve overall ecosystem health.

One of the most fascinating applications of AI in biodiversity is in habitat monitoring. AI-powered drones and sensors can track various species' movements and interactions within agricultural land. This data can then be analyzed to create models that predict how different farming practices will impact local wildlife and plant species. In some cases, AI has already been used to identify "biodiversity hotspots" within farmlands where natural flora and fauna thrive. Protecting and nurturing these spots can lead to a more balanced and healthier ecosystem.

In addition to monitoring and analysis, AI also plays a crucial role in conservation planning. Advanced simulations can predict the outcomes of different conservation strategies, helping farmers make informed decisions. These simulations can incorporate a myriad of variables, including climate change projections, water availability, and land use patterns. By simulating the long-term impacts of different farming practices, AI can guide farmers toward more sustainable and conservation-friendly methods.

Economic viability is often the major concern when it comes to adopting new agricultural practices. However, AI-driven models can also analyze the economic benefits of diverse cropping systems versus traditional monocultures. These models can project potential savings in terms of reduced need for chemical inputs like fertilizers and pesticides, which are often associated with monocultures. Additionally, diversified farms can become less vulnerable to market fluctuations related to single crops, thereby offering a more stable income stream for farmers.

The significance of AI in promoting biodiversity extends to pest and disease management as well. Traditional pest control methods often involve broad-spectrum chemicals that can harm non-target species and wreak havoc on local ecosystems. In contrast, AI systems can identify pests and diseases with pinpoint accuracy, suggesting targeted interventions that minimize collateral damage. For example, precision spraying technologies can deliver chemicals only to affected areas, leaving the rest of the ecosystem untouched.

Water management is another domain where AI's impact on biodiversity is profoundly felt. Over-irrigation and the misuse of water resources can severely disrupt local hydrological cycles, affecting aquatic habitats and terrestrial ecosystems. AI tools can optimize irrigation schedules based on real-time soil moisture data and weather forecasts, ensuring water is used efficiently. By maintaining natural water cycles, farmers can contribute to healthier local waterways, benefiting both plant and animal species.

Promoting biodiversity and conservation goes hand-in-hand with protecting pollinators, essential agents in the reproductive stages of many crops. Pollinators like bees and butterflies are increasingly threatened by habitat loss and pesticide use. AI can help by tracking pollinator populations and identifying plants that support them. With this data, farmers can create pollinator-friendly zones within their fields, ensuring these vital species receive the resources they need to thrive.

Soil health is foundational to any conservation effort. Degraded soils lose their ability to support diverse plant life, turning into barren landscapes over time. AI-driven soil analysis offers a detailed understanding of soil composition and health, enabling precise interventions. By recommending soil amendments and crop rotations that enhance soil biodiversity, AI helps build soil resilience. Healthy soils, rich in organic matter and microorganisms, support a wider variety of plants and, subsequently, a greater diversity of other organisms.

Furthermore, artificial intelligence facilitates community engagement in biodiversity and conservation efforts. By providing accessible data and visualizations, AI can help make the importance of these initiatives clear to local communities and stakeholders. Engaging local communities ensures that conservation efforts are not only top-down mandates but collaborative endeavors. When farmers and local stakeholders see the tangible benefits—such as improved crop yields, reduced pest infestations, and enhanced water quality—they are more likely to participate actively in conservation initiatives.

From a broader perspective, AI enables a shift from reactive to proactive conservation strategies. Instead of responding to crises like pest outbreaks or water shortages, farmers can anticipate and mitigate these issues before they become severe. Predictive analytics powered by AI allow for early warnings about potential threats to biodiversity, giving farmers the lead time needed to take corrective actions. This proactive approach aligns well with sustainable farming principles, ensuring long-term ecological and economic benefits.

Integrating AI into farming practices also means safeguarding rare and endangered species. For instance, specific algorithms can identify and track rare plant or animal species within farmland. This data can then be used to create conservation zones or corridors that protect these species while allowing farming activities to continue in a sustainable manner. By balancing agricultural productivity with the preservation of rare species, AI offers a harmonious solution to modern farming challenges.

Promoting biodiversity and conservation isn't just about protecting nature; it's about creating a resilient and sustainable agricultural system for future generations. As AI continues to evolve, its potential to drive positive environmental outcomes will only grow. The integration of AI in agriculture offers a new frontier for conserving our natural world while meeting the food needs of a growing global population. This synergy of technology and sustainability could very well be the cornerstone of the next green revolution.

In conclusion, while traditional farming heavily relied on intensive, often detrimental practices, AI presents an opportunity to rethink and redesign the agricultural paradigm. AI acts as both a microscope and a telescope, providing granular insights into present conditions while forecasting future scenarios. This dual capability makes it an invaluable tool in promoting biodiversity and conservation. By embracing AI, farmers around the world can adopt more sustainable practices that not only boost productivity but also contribute to a healthier, more diverse ecosystem. The ripple effects of this transformation go far beyond the farm; they reach into the very fabric of our natural world, forging a future where agriculture and conservation thrive in unison.

Chapter 15:
Economic Benefits of AI in Agriculture

The integration of AI in agriculture offers transformative economic benefits, ranging from significant cost reduction to enhanced operational efficiency. By automating labor-intensive tasks, AI reduces the need for manual labor, thereby decreasing operational costs and improving productivity. Precision farming techniques, driven by AI, enable farmers to optimize resource use, such as water, fertilizers, and pesticides, leading to lower input costs and higher yields. Additionally, AI-powered predictive analytics help in timely decision-making, reducing crop failures and financial risks. These efficiencies not only bolster farm profitability but also make farming more sustainable, paving the way for a resilient agricultural economy. Embracing these technologies isn't just about staying competitive; it's about securing the future of farming in an increasingly resource-constrained world.

Cost Reduction and Efficiency Gains

Artificial intelligence (AI) is revolutionizing the agricultural sector by significantly reducing costs and enhancing efficiency. This transformation is enabling farmers to do more with less, maximizing resources and streamlining processes. One of the primary ways AI achieves this is through automation. Autonomous machinery like AI-powered tractors and harvesters can operate around the clock, reducing labor costs and minimizing human error.

Another key aspect is the predictive capabilities of AI. By analyzing vast amounts of data from various sources, AI systems can predict equipment failures before they happen. This kind of predictive maintenance reduces downtime and avoids the high costs associated with unexpected repairs. Farmers can schedule maintenance during non-peak times, ensuring that equipment is always in optimal condition when needed most.

Furthermore, AI is instrumental in optimizing planting and harvesting schedules. Traditional farming methods rely heavily on experience and intuition, which can lead to inefficiencies and waste. AI systems, on the other hand, use real-time data and advanced algorithms to determine the best times for planting and harvesting, taking into account factors like weather patterns, soil health, and crop conditions. This optimization improves yield and reduces waste, translating directly into cost savings.

Water management is another area where AI can significantly cut costs and increase efficiency. Traditional irrigation systems can be wasteful, with water ending up in areas where it's not needed. AI-driven solutions, such as precision irrigation, ensure that water is delivered only where and when it's needed. This targeted approach not only conserves water but also reduces energy costs associated with pumping and distributing it.

AI also plays a crucial role in crop monitoring and management. Drones and satellite imagery equipped with AI can provide real-time surveillance of large farm areas, identifying issues like pest infestations or nutrient deficiencies before they become widespread problems. Early detection allows for timely interventions, reducing the amount of pesticides and fertilizers needed. This not only cuts costs but also promotes sustainable farming practices.

The benefits of AI extend to supply chain optimization as well. Efficient logistics and distribution are crucial for getting products to market quickly and at the lowest cost. AI can analyze data such as traffic patterns, weather conditions, and market demand to optimize delivery routes and schedules. This leads to reduced transportation costs and faster delivery times, ensuring that produce remains fresh and commands the best prices.

Labor is often one of the largest expenses in farming, and AI is helping to alleviate some of these costs through automation and efficiency gains. Automated feeding and milking systems for livestock, for example, reduce the need for manual labor and ensure that animals are fed and milked consistently. This leads to healthier animals, higher yields, and lower long-term costs.

Moreover, AI enhances decision-making by providing farmers with actionable insights derived from big data. Tools that analyze market trends, weather forecasts, and crop health can help farmers make informed decisions that optimize resource use and increase profitability. Predictive analytics can also assist in negotiating better prices and contracts, further boosting economic gains.

The initial investment in AI technology can be substantial, but the long-term savings and efficiency gains often far outweigh the costs. Governments and financial institutions are starting to recognize this and are offering incentives and funding opportunities to help farmers adopt AI technologies. These initiatives can help mitigate the financial burden of the initial investment, making AI more accessible to a broader range of farmers.

It's crucial to understand that the economic benefits of AI in agriculture aren't just limited to large-scale operations. Smallholder farmers also stand to gain significantly. Affordable AI tools tailored for smaller farms can help them achieve similar efficiency gains and cost reductions. This democratization of technology ensures that the benefits of AI are felt across the agricultural spectrum, contributing to a more sustainable and equitable agricultural system.

In summary, AI is driving a paradigm shift in agriculture by reducing costs and improving efficiency across various facets of farming. From autonomous machinery and predictive maintenance to precision irrigation and supply chain optimization, the economic benefits are immense. These advancements not only enhance profitability but also promote sustainable practices, ensuring that the agricultural sector is well-equipped to meet the challenges of the future.

Economic Challenges and Considerations

The economic benefits of AI in agriculture are vast, but they come with their own set of challenges and considerations. First and foremost are the initial costs associated with integrating AI technologies. It's no secret that implementing AI systems can be expensive. From purchasing advanced sensors and machinery to investing in specialized software, the startup costs can be prohibitively high for many small and medium-sized farms. Although these investments often pay off in the long term, the initial financial barrier can be a significant deterrent.

Another critical consideration is the scalability of AI technologies. While large, industrial farms might have the resources to adopt and benefit from AI innovations, smaller farms might struggle with these transitions. There's an uneven playing field when it comes to the adoption of AI—a gap often referred to as the "digital divide." This disparity can lead to an even greater economic gap within the agricultural sector, disadvantaging family-run and small-scale farms.

Moreover, the integration of AI technologies requires a skilled workforce to manage and maintain these systems. This demand for new skills brings about a need for substantial investments in training and education. Farmers and their teams must be familiar with data analytics, machine learning models, and the operational nuances of autonomous machinery. This educational shift doesn't just involve time and money, but also a cultural change in an industry deeply rooted in traditional practices.

Then there's the issue of data privacy and security. AI in agriculture relies heavily on data collection, employing sensors and drones to gather information about soil conditions, crop health, weather patterns, and more. Ensuring that this data is securely stored and only accessible by authorized personnel is paramount. Any data breaches could result in serious financial losses, eroding trust in these technologies. Adopting robust cybersecurity measures, therefore, becomes an indispensable component of integrating AI, adding another layer of cost and complexity.

Transitioning to AI-driven practices also implies a period of adjustment during which productivity may temporarily decline. The learning curve associated with new technologies can lead to initial inefficiencies and disruptions in farm operations. Farmers may need to reconfigure their existing workflows and practices to align with the capabilities and requirements of AI technologies. This transition phase can be particularly challenging, creating short-term economic pressures even as long-term gains loom on the horizon.

Furthermore, the economic advantages offered by AI are often influenced by external factors such as government regulations and policy frameworks. Regulatory environments may not keep pace with rapid advancements in AI, leading to ambiguities and uncertainties. For example, policies related to data ownership, liability in case of malfunctioning autonomous machinery, and standards for AI applications can vary widely, affecting the smooth economic integration of these technologies in agriculture.

Another economic challenge is the risk of redundancy and job displacement. AI technologies are poised to automate a range of tasks, from planting to harvesting, which could render certain manual jobs obsolete. This shift can lead to economic displacement for many workers in the agricultural sector. While the rise of AI could open new roles requiring different skill sets, there is an immediate need to balance automation with human employment, safeguarding vulnerable livelihoods.

International market dynamics also play a role in shaping the economic landscape for AI in agriculture. Countries with broader access to AI technologies may gain competitive advantages, potentially disrupting global agricultural trade patterns. Nations lagging in AI adoption could suffer economic setbacks, leading to increased import dependencies and possible food security issues. This uneven adoption could further strain international relations and trade agreements.

Lastly, the rapid pace of technological advancement poses an economic challenge. AI systems and tools are continually evolving, which can lead to issues with obsolescence. Farms may invest heavily in current technologies only to find them outdated a few years after implementation. This need for continuous updates and upgrades can add a recurring financial burden, requiring farmers to constantly reinvest in newer technologies to remain competitive.

In conclusion, while the economic potential of AI in agriculture is immense, various challenges and considerations need careful navigation. Upfront costs, scalability issues, educational demands, data privacy concerns, transitional inefficiencies, regulatory uncertainties, job displacement risks, global market implications, and rapid technological changes are all significant factors. Successfully addressing these challenges is vital for realizing the full economic benefits that AI promises for the agricultural sector. As we continue to explore and innovate, these considerations will shape the trajectory of AI's integration into farming practices, driving us towards a more technologically advanced and economically viable future in agriculture.

Chapter 16:
Policy and Regulation

As artificial intelligence (AI) continues to revolutionize agriculture, understanding the nuances of policy and regulation becomes crucial. Governments worldwide are grappling with the pace at which AI technologies are advancing, striving to create frameworks that foster innovation while safeguarding ethical considerations and data privacy. In the agricultural sector, these regulations need to balance the benefits of AI, such as increased efficiency and sustainability, with potential risks, including data misuse and job displacement. Navigating AI-related agricultural policies requires collective effort from policymakers, farmers, technologists, and stakeholders, ensuring that regulations are not just reactive but also forward-looking. Transparent guidelines and robust legal frameworks can help mitigate risks, encourage responsible AI usage, and ultimately, drive the next green revolution in a manner that is inclusive and equitable for all.

Navigating AI-related Agricultural Policies

Navigating the evolving landscape of AI-related agricultural policies can feel like venturing into uncharted territory. These policies are vital as they define the framework within which AI technologies operate in the agricultural sector. As governments and regulatory bodies across the globe grapple with the implications of AI, it becomes critical for farmers, tech developers, and investors to stay abreast of these regulations. Understanding the legal landscape ensures compliance and helps leverage AI for maximum productivity while safeguarding ethical standards and data privacy.

In recent years, several countries have recognized the potential of AI in transforming agriculture and have started developing specific policies to encourage its adoption. These policies often focus on promoting research and development, subsidizing AI technologies, and providing guidelines for ethical AI use. For example, the U.S. Department of Agriculture (USDA) has been actively supporting AI innovations through grants and partnerships designed to foster modern farming technologies. Similarly, the European Union's Common Agricultural Policy (CAP) includes provisions to support digital farming initiatives, emphasizing sustainability and technological adoption.

However, the fast-paced evolution of AI technologies requires policies that are adaptable and forward-thinking. Static regulations may quickly become obsolete as new AI applications emerge. Hence, regulatory bodies need to strike a balance between providing clear guidelines and allowing the flexibility necessary for innovation. For instance, sandbox approaches, where new technologies can be tested under regulatory supervision, have proven effective in other sectors and could be beneficial in agriculture as well.

Data privacy remains one of the most critical issues in AI-related agricultural policies. With AI systems relying heavily on data to provide accurate predictions and actionable insights, the collection, storage, and use of data must adhere to stringent privacy standards. Policies such as the General Data Protection Regulation (GDPR) in Europe mandate rigorous data protection measures. Farmers and AI developers must ensure compliance to avoid hefty penalties and maintain public trust. Moreover, transparent data practices can encourage more farmers to adopt AI tools, knowing their data is secure and used responsibly.

Ethical considerations intersect closely with policy development. The deployment of AI in agriculture should not exacerbate existing inequalities. Policymakers must consider the socioeconomic impacts of AI technologies, ensuring they benefit small-scale farmers as much as large agricultural enterprises. Providing equal access to training and resources can democratize the advantages of AI, preventing a technological divide from widening. Policymakers should promote inclusive growth by implementing subsidies and financial incentives tailored to smaller farming operations.

When discussing policy, we cannot overlook the environmental regulations dictating AI's role in sustainable agriculture. AI's prowess in optimizing resource use, such as water and fertilizers, aligns well with global environmental goals. Policies aimed at reducing the agricultural carbon footprint are increasingly incorporating AI as a key component. For example, programs that incentivize precision farming techniques can lead to more efficient farming practices, conserving resources and reducing greenhouse gas emissions.

Collaboration is another cornerstone in the realm of AI-related agricultural policies. Governments, private sector stakeholders, and academic institutions need to work together to create a coherent policy environment. Initiatives such as public-private partnerships can accelerate the adoption of AI in agriculture by pooling resources and expertise. Moreover, international cooperation can harmonize regulations, making it easier for AI technologies to cross borders and benefit a wider audience. Shared frameworks and standards facilitate the global exchange of data, practices, and innovations.

Additionally, future AI-related agricultural policies should consider the integration of AI education into agricultural curriculums. By equipping the next generation of farmers and agricultural professionals with AI expertise, we ensure a smoother transition to advanced farming methods. Educational policies that support AI training programs, workshops, and research initiatives can create a knowledgeable workforce ready to harness AI's potential. Investments in educational resources and institutions will yield long-term benefits, fostering an innovative and resilient agricultural sector.

Challenges abound when formulating AI-related agricultural policies, particularly in developing nations where resources and infrastructure might be lacking. Policymakers in these regions must prioritize creating an enabling environment that supports AI development and adoption. This could involve investment in digital infrastructure, offering subsidies for AI technologies, and fostering collaborations with international partners. Ensuring accessibility for all segments of society will be crucial in maintaining equitable growth and maximizing the benefits of AI.

Moreover, regulators must continually update policies to keep pace with AI advancements. Rapid technological changes require agility and a willingness to adapt. Periodic reviews and stakeholder consultations can ensure that policies remain relevant and effective. Involving farmers, tech developers, and researchers in policy-making processes can provide valuable insights, ensuring that regulations are grounded in practical realities and scientific advancements.

In essence, effective AI-related agricultural policies must balance innovation with responsibility. They should facilitate the adoption and development of AI technologies, ensure ethical and equitable access, prioritize data privacy, and support environmental sustainability. By navigating this complex policy landscape, stakeholders can unlock AI's full potential, driving the next green revolution and fostering a sustainable, productive, and technologically advanced agricultural sector.

Ethical Considerations and Data Privacy

As artificial intelligence continues to revolutionize agriculture, the potential for ethical dilemmas and data privacy concerns intensifies. From crop monitoring to livestock management, AI technologies are increasingly collecting and analyzing vast amounts of data. This collection poses significant questions about ownership, consent, and the ethical use of information.

One of the core ethical considerations is the matter of data ownership. When sensors and AI tools collect data from a farm, it's essential to establish who owns this data. Is it the farmers who own their land and crops, or the companies providing these sophisticated technologies? It's a gray area that many current regulations don't fully address. This lack of clarity can lead to exploitation, where the value generated from the data does not return to the original owners— potentially depriving farmers of crucial benefits.

Furthermore, consent is another complex issue. Farmers must be fully aware of and agree to the data collection processes occurring on their properties. Transparent policies should be implemented to ensure that farmers understand what data is being collected, how it's being used, and who has access to it. To build trust, these policies should be written in straightforward language, avoiding technical jargon that might confuse end-users.

Beyond ownership and consent, there are broader ethical implications around equity and access. While AI technologies have the potential to dramatically increase agricultural productivity, they often require significant investment. This requirement can create discrepancies between wealthy, large-scale farms and smaller, less-resourced operations, potentially widening the gap between them. Ensuring equitable access to these technologies is crucial for a fair and inclusive growth in the agricultural sector.

Data privacy is intrinsically linked to these ethical challenges. With the increasing deployment of AI technologies comes the risk of data breaches and misuse. Agricultural data, which may include sensitive information about land usage, crop yields, and livestock health, is invaluable not only to individual farmers but also to competitors and external malicious actors. Thus, robust cybersecurity measures must be established to safeguard this information.

Regulations can play a pivotal role in addressing these concerns. Policymakers worldwide are grappling with the dual challenge of fostering innovation while ensuring ethical conduct and data privacy. Given the rapid pace at which AI technology evolves, regulations must be adaptive and forward-thinking. A comprehensive framework that includes standardized guidelines on data collection, storage, and usage is vital.

An important aspect of these regulations should be ensuring data anonymization, which involves stripping data of identifying markers to protect individual identities while maintaining its utility for analysis. Methods like differential privacy can also be employed to introduce "noise" into datasets, further protecting sensitive information without compromising the data's overall value.

Ethical AI usage also involves considering the long-term impacts of AI decisions. For example, when AI-driven recommendations on pesticide use are given, the environmental and health implications should be thoroughly evaluated. Ethically aligned AI should promote sustainable practices, thereby supporting long-term agricultural health and environmental conservation.

Moreover, the principle of fairness must be embedded in AI systems used in agriculture. This means ensuring that AI algorithms do not inadvertently introduce or amplify biases, such as those based on geographical or socio-economic factors. Bias in AI could lead to unequal recommendations and opportunities, disadvantaging certain groups of farmers. Engaging diverse datasets and implementing fairness audits can help mitigate this risk.

Another ethical dimension is the transparency of AI systems. Farmers and stakeholders should know how an AI system arrived at a particular recommendation or decision. This transparency builds trust and allows users to contest and understand decisions that might otherwise seem arbitrary. Explainable AI, which focuses on making AI decision-making processes understandable to humans, is crucial in this context.

Education plays a significant role in addressing ethical and data privacy concerns. Training programs for farmers and agricultural professionals should include modules on the ethical use of AI and data privacy. By equipping them with the necessary knowledge, we empower them to make informed decisions and advocate for their rights effectively.

International cooperation is also necessary to address these challenges effectively. Agricultural practices and regulations differ vastly around the world, and an international consensus on best practices for ethical AI and data privacy could go a long way in ensuring global standards. Platforms for dialogue among countries and stakeholders can facilitate the sharing of knowledge, leading to more cohesive and effective regulations.

Ultimately, the ethical considerations and data privacy issues in AI agriculture are not merely technical challenges; they are human ones. They compel us to consider the broader impact of our technological advancements on individuals, communities, and the environment. Thus, as we navigate this new frontier, maintaining a balanced approach that prioritizes ethical integrity and robust data privacy safeguards is not just desirable—it is imperative.

The onus also lies on tech companies to act responsibly. They should implement stringent ethical guidelines internally and engage with the agricultural communities they serve. Collaborative efforts, where farmers, tech companies, and policymakers work together to create and uphold these ethical standards, are necessary to foster a sustainable and equitable agricultural future.

In conclusion, the journey toward integrating AI in agriculture is fraught with ethical considerations and data privacy challenges. By focusing on the principles of ownership, consent, equity, security, transparency, and education, we can address these issues head-on. Thus, we pave the way for a technologically advanced yet ethically responsible agricultural paradigm that benefits all stakeholders equally.

Chapter 17:
Global Case Studies

From the high-tech rice fields of Japan to the expansive wheat farms in Canada, AI is revolutionizing agriculture on a global scale. Whether it's using drones for real-time crop monitoring in Brazil or utilizing AI-enabled soil sensors in Kenya, these global case studies showcase the transformative power of technology in diverse environments. In India, AI-driven predictive analytics have significantly reduced crop losses, while in the Netherlands, automated greenhouses are optimizing yield and resource use. Each case study offers unique lessons, illustrating how AI adapts to varying climates, cultures, and agricultural practices. These success stories not only demonstrate the potential of AI to enhance productivity and sustainability but also underscore the importance of tailoring solutions to local conditions.

Successful Implementations Around the World

As we dive into the diverse case studies of AI's implementation in agriculture, it's clear that this technology is not a one-size-fits-all solution. Different regions, each with their unique climates and agricultural practices, have found innovative ways to harness AI to address their specific challenges. From the rice paddies of Asia to the vast wheat fields of North America, AI is proving to be a transformative force.

In Japan, for instance, AI-driven farming is not just an experiment; it's a necessity. With a declining farming population and limited arable land, the country needed a way to maintain and even increase productivity. Farmers have turned to AI-powered drones and robots for planting, maintaining, and harvesting crops. These machines are equipped with sensors and cameras that analyze soil quality, monitor crop health, and even predict the best times for planting and watering. Japanese farmers have reported a significant increase in yields and a reduction in labor costs.

Moving to Africa, AI has been instrumental in addressing some of the continent's most pressing agricultural challenges. In Kenya, for example, an AI application called PlantVillage Nuru is making waves. This app helps farmers identify crop diseases through images captured on their smartphones. Using machine learning algorithms, the app provides a diagnosis and suggests treatment options. This has been revolutionary, as millions of farmers can now take preventative action before a small issue turns into a devastating problem. The result? Decreased crop loss and improved food security.

India presents another fascinating case. With its diverse climate and an economy heavily reliant on agriculture, the integration of AI has had a profound impact. Companies like CropIn Technology have developed AI-based platforms that offer real-time advisory services on weather, soil health, and crop management. Farmers can access this information via mobile apps, enabling them to make informed decisions. This technology has been particularly beneficial to small and marginal farmers, who traditionally have limited access to agricultural expertise. The increased productivity and efficiency are helping drive a new agricultural renaissance in the country.

In the Netherlands, the integration of AI into sustainable farming practices is nothing short of a technological marvel. The country is renowned for its advanced greenhouse technologies, and AI has taken these to new heights. Dutch farmers use AI to control every aspect of the greenhouse environment, from temperature and humidity to CO_2 levels and lighting. Machine learning algorithms continuously analyze this data to optimize growing conditions. The outcome? Exceptionally high yields and resource-efficient farming, serving as exemplary models for urban agriculture systems globally.

Across the Atlantic, AI's role in the agricultural sector of the United States cannot be understated. American farmers are leveraging AI to combat challenges ranging from unpredictable weather patterns to labor shortages. One standout example is The Climate Corporation, which utilizes machine learning to provide precise weather analytics and planting recommendations. This enables farmers to better plan their planting schedules and irrigation strategies, reducing waste and increasing profitability. Additionally, AI-driven machinery, like autonomous tractors and robotic harvesters, contribute significantly to labor efficiency.

In Brazil, the world's largest producer of coffee, AI is helping to elevate the industry to new standards. Agricultural firms are using AI to monitor coffee crops for diseases, optimize harvest schedules, and even improve the quality of the beans. Drones equipped with image sensors send real-time data to AI systems, which analyze the health of the crops and predict the best times for harvesting. This meticulous attention to detail ensures that the coffee beans harvested are of premium quality, enhancing Brazil's reputation as a top coffee exporter.

Australia's vast and varied landscapes present unique agricultural challenges, which AI is helping to tackle. The country's beef industry, in particular, has seen significant advancements through AI-enabled livestock monitoring systems. These systems use cameras and sensors to monitor the health, movement, and behavior of cattle, providing farmers with valuable insights they previously lacked. Such technologies not only improve animal welfare but also enhance productivity and profitability. By catching diseases early and optimizing feeding schedules, Australian farmers are seeing improved herd conditions and higher quality beef.

In Israel, innovation and agriculture go hand-in-hand, and AI has become a cornerstone of their success. Israeli startups are pioneering AI technologies to maximize water efficiency through precision irrigation systems. Technologies like CropX use soil sensors and AI to provide real-time recommendations on irrigation needs. This is especially crucial in a country where water is a scarce resource. The precise application of water ensures that crops receive the optimal amount, reducing waste and promoting sustainable farming practices. Such advancements have made Israel a global leader in agricultural technology.

Exploring China, the rapid and large-scale implementation of AI in agriculture is illustrative of the country's commitment to food security and technological advancement. Chinese farms are increasingly adopting AI-driven platforms for soil analysis, crop disease detection, and yield prediction. Companies like XAG are deploying drones and robots to assist in planting and pest control. The agricultural outputs have significantly benefited, with a noted increase in efficiency and productivity across various crops. Chinese tech giants are also investing heavily in AI research to push the boundaries of what's possible in agriculture.

While these examples demonstrate the remarkable success of AI implementations in agriculture, it's essential to recognize the broader implications. The efficiency gains, resource optimizations, and productivity increases are not merely regional wins; they contribute to global food security and sustainability. AI-driven solutions in agriculture can mitigate some of the most critical challenges we face today, from climate change to hunger. Each successful implementation offers lessons that can be shared and adapted across different regions, promoting a collaborative approach to agricultural innovation.

In summary, the adoption of AI in agriculture is shaping up to be a powerful catalyst for the next green revolution. From the advanced greenhouses of the Netherlands to the innovative irrigation systems in Israel, AI is making agriculture more efficient, sustainable, and productive. These successful implementations around the world not only provide tangible benefits to local farmers but also pave the way for a more food-secure and environmentally conscious future. By learning from these examples, we can continue to innovate and push the boundaries of what's possible in agriculture, ensuring that AI's transformative power is harnessed for the greater good of humanity and the planet.

The journey of AI in agriculture is just beginning, but its impact is already profound. As more regions adopt these technologies, the success stories will multiply, offering a beacon of hope and inspiration for farmers and technologists worldwide. The mission is clear: to support sustainable agriculture and ensure our food systems are resilient against future challenges. And AI is proving to be an invaluable partner in this mission, carving a path towards a brighter, more sustainable future.

Lessons Learned from Diverse Climates and Cultures

Diverse climates and cultures around the globe present a rich tapestry of lessons for those at the intersection of artificial intelligence (AI) and agriculture. Each region brings forth unique challenges and opportunities, paving the way for innovations that might otherwise remain undiscovered. The heterogeneous nature of the world's agricultural landscapes makes it imperative for AI solutions to be adaptive and context-specific.

Let's start with the understanding that climatic diversity demands tailored solutions. In regions with arid climates, such as the Middle East and parts of Africa, water scarcity is a significant challenge. AI-driven solutions for efficient irrigation have made remarkable strides here. For instance, AI algorithms can analyze soil moisture data in real-time, enabling farmers to optimize water usage. This has not only enhanced crop yields but also conserved precious water resources. The lesson learned is clear: AI applications must be responsive to the specific environmental conditions they are deployed in.

On the other hand, regions like Southeast Asia, with their tropical climates, have their own set of challenges. High humidity levels and persistent rainfall can lead to the rapid spread of diseases among crops. AI technologies, particularly those involved in disease detection and prevention, have been a game-changer. By utilizing machine learning models trained on vast datasets of plant health indicators, farmers can identify diseases in their early stages and take preventive measures. This proactive approach has reduced crop loss and increased productivity. Here, the importance of swiftly adapting AI models to local conditions is underscored.

Moving to culturally rich and agriculturally significant regions like Latin America, especially Brazil and Argentina, lessons emerge on integrating traditional farming knowledge with AI. These countries have vast agricultural areas and a deep-rooted history of farming practices passed down through generations. The fusion of this traditional knowledge with AI insights has proven to be incredibly effective. For example, AI can predict the best planting and harvesting times by analyzing historical weather patterns and current climatic conditions. Farmers, relying on their ancestral knowledge, can further validate and fine-tune these predictions. This synergy between AI and age-old wisdom highlights the necessity of incorporating local knowledge into technological solutions.

Moreover, regions with diverse terrains, like India, offer additional insights. India's agrarian landscape includes everything from arid deserts and plains to lush, fertile valleys. This diversity calls for highly localized AI solutions. Smallholder farmers, who represent the backbone of Indian agriculture, benefit immensely from mobile-based AI applications that provide real-time advice on crop management. The locally customized recommendations have empowered these farmers to make informed decisions, significantly improving their livelihoods. The lesson here is the critical role of accessibility and localization in making AI effective for small-scale farmers.

In European countries with well-established agricultural sectors, such as the Netherlands, the focus has often been on maximizing efficiency and sustainability through AI. Smart greenhouses equipped with AI technologies are used to monitor and adjust growth conditions, optimizing factors like light, temperature, and humidity for different crops. This has significantly enhanced both the quality and quantity of produce. The takeaway from these advanced regions is the potential of AI to push the boundaries of agricultural productivity while maintaining sustainability.

Another fascinating aspect is the emphasis on community engagement and social factors when deploying AI in agriculture. In countries like Kenya and Ethiopia, farmer cooperatives and community-based organizations have been instrumental in the successful implementation of AI technologies. These groups act as intermediaries, ensuring that AI tools are not only disseminated but also effectively utilized. This social approach to technology deployment ensures better adoption rates and more significant impact. The key lesson here is the importance of community involvement and social structures in driving technological success.

Meanwhile, countries experiencing rapid technological advancements, like China, have demonstrated the scalability potential of AI in agriculture. Large-scale farms in Chinese provinces utilize AI for various aspects, from crop monitoring and pest control to logistics and supply chain optimization. The ability to integrate multiple AI applications into a comprehensive, scalable system has resulted in substantial efficiency gains and cost reductions. This scalability showcases how a coordinated approach can amplify the benefits of AI across vast agricultural landscapes.

Let's not overlook the unique agricultural practices found in island nations like Japan. In these regions, space constraints pose a significant challenge. AI-powered vertical farming and hydroponics systems have been breakthroughs, allowing for efficient use of limited space. These systems use AI to monitor plant growth conditions and automate nutrient delivery, ensuring optimal growth. The innovation here lies in the adaptation of AI to maximize agricultural output in confined spaces, demonstrating AI's flexibility and creative potential.

From these varied experiences, one overarching lesson stands out: adaptability is key. The diverse climates and cultures across the globe teach us that there is no one-size-fits-all solution in agriculture. AI technologies must be designed with the flexibility to accommodate a wide range of environmental conditions and cultural contexts. This adaptability ensures that the benefits of AI reach every corner of the agricultural world, irrespective of the climatic or cultural nuances present.

Equally important is the layer of continuous learning and feedback. As AI systems are deployed across the globe, they gather a wealth of data from different environments and practices. This data, when analyzed and integrated, contributes to the refinement and evolution of AI technologies. The dynamic nature of this knowledge loop ensures that AI solutions become increasingly robust and versatile over time.

Lastly, the global case studies reveal that collaboration and knowledge sharing are critical in advancing AI-driven agriculture. Countries and communities that share their successes and challenges enable a collective learning process. Whether through international conferences, research collaborations, or cross-border partnerships, the exchange of ideas accelerates innovation and adaptation. The synthesis of global experiences fosters a rich environment for the continuous improvement of AI applications in agriculture.

In conclusion, the lessons learned from diverse climates and cultures emphasize that the transformative power of AI in agriculture lies in its adaptability and integration with local knowledge and conditions. By embracing these lessons, we can harness AI's potential to drive the next green revolution, ensuring food security and sustainability for future generations.

Chapter 18:
AI in Small-Scale Farming

AI in small-scale farming is revolutionizing how smallholder farmers operate, offering powerful tools and technologies that were once reserved for larger operations. With advancements like AI-driven soil sensors, predictive analytics, and automated irrigation systems, even modest farms can achieve significant boosts in efficiency and productivity. This democratization of technology empowers small farmers to make data-driven decisions, improving crop yields and resource management while reducing costs. As AI becomes more accessible and affordable, its potential to promote rural development and sustainable farming practices cannot be overstated. Indeed, AI stands poised to transform the backbone of agriculture—small-scale farms—into hubs of innovation and productivity, fostering a resilient and equitable agricultural future for all.

Tools and Technologies for Small Farmers

Small-scale farmers often operate under significant constraints—limited land, finances, and access to advanced technologies. Yet, the integration of artificial intelligence (AI) and machine learning (ML) tools holds incredible promise for transforming these limitations into opportunities. While large-scale operations have long benefited from advanced technologies, innovations tailored for small farmers are now emerging at an unprecedented rate, democratizing access to state-of-the-art resources.

141

AI-powered tools can revolutionize tasks that small farmers often find labor-intensive. Take soil analysis, for instance. Traditional soil testing requires sending samples to a lab, a procedure both time-consuming and costly. Modern AI-driven soil sensors, however, can analyze soil conditions in real-time, providing actionable insights on nutrient levels, pH balance, and moisture. This instant feedback empowers farmers to make timely decisions, optimizing input use, and improving crop yields.

Moreover, AI is transforming crop monitoring and management into a science of precision. Drones equipped with AI software can fly over fields, capturing high-resolution images and data that can detect the onset of disease, stress, or pest infestations long before they are visible to the naked eye. These AI algorithms can differentiate between various stresses—whether they are due to lack of water, nutrient deficiency, or disease—allowing for precise interventions. For small farmers, such technologies mean less waste, lower costs, and healthier crops.

One can't overlook the development of mobile applications designed to bring the power of AI to the fingertips of small farmers. Apps like Plantix and FarmLogs use machine learning algorithms to provide farmers with weather forecasts, crop disease diagnostics, and pest management solutions. They offer tailored advice based on real-time data and even historical comparisons from similar farms in the region. The ability to access such powerful tools via a smartphone opens up a world of possibilities for small-scale farmers, making advanced agricultural knowledge more accessible than ever before.

Remote sensing technology is yet another frontier where AI is making a significant impact. Small farmers often lack the resources to conduct detailed land surveys. AI-enabled satellite imagery and remote sensors can fill this gap, offering data on everything from soil health to crop growth stages and water usage. By leveraging remote sensing, farmers can manage resources more effectively, optimize planting schedules, and enhance their farm's overall productivity.

Integrating AI solutions into irrigation systems is another game-changer. Water management is crucial, especially for small farms that rely on precise irrigation to ensure crop health. AI algorithms can analyze weather data, soil moisture levels, and crop requirements to automate irrigation systems. These smart systems ensure that crops receive the exact amount of water needed, reducing waste and conserving a precious resource. This not only leads to better crop yields but also significant cost savings.

Agricultural robotics is no longer a domain exclusive to large-scale operations. Smaller, affordable robots are being developed to cater to the unique needs of small farms. These robots can handle tasks like planting seeds, weeding, and harvesting with high accuracy and efficiency. Autonomous weeders, for instance, use machine vision to identify weeds and remove them without harming crops. Such innovations reduce the need for manual labor and chemical herbicides, promoting more sustainable farming practices.

AI has also made strides in the realm of pest control. Predictive analytics can forecast pest infestations by analyzing data from various sources, including satellite imagery, historical weather patterns, and local reports. These predictions allow farmers to take preemptive measures, significantly reducing crop damage and minimizing the use of pesticides. Small farms, which are often more vulnerable to pest outbreaks, stand to benefit immensely from these predictive capabilities.

Support systems like virtual assistants and chatbots are increasingly common. Tailored for small farmers, these AI-driven tools can answer questions, offer farming advice, and provide reminders for essential tasks such as planting, fertilizing, and harvesting. Imagine a scenario where a farmer can simply ask a virtual assistant about the best time to plant a particular crop based on predictive analytics and receive an accurate, customized response. This level of support could dramatically enhance farm management and productivity.

Supply chain optimization is another area where AI is making its mark. Traditionally, small farmers have struggled to navigate complex supply chains, often settling for lower prices due to a lack of market information. AI-driven platforms can now provide real-time data on market prices and demand trends, enabling farmers to make informed decisions about when and where to sell their produce. This level of insight helps farmers maximize their profits and reduces post-harvest losses.

The innovation extends to financial tools as well. AI can provide small farmers with credit scores based on alternative data sources, such as mobile transaction records and social media activity, circumventing traditional banking methods that require extensive documentation and collateral. These AI-driven credit assessments open up new avenues for loans and investment, particularly beneficial for farmers in regions with limited access to traditional financial services.

Collaborative platforms enabled by AI can foster community-based farming practices. These platforms connect small farmers with each other, facilitating the sharing of equipment, labor, and knowledge. By pooling resources, farmers can reduce costs and access technologies that might otherwise be out of reach. AI helps in optimizing the allocation of shared resources, ensuring fairness and efficiency.

While the promise of AI in small-scale farming is significant, it's essential to acknowledge the barriers to adoption, particularly in developing regions. Limited internet connectivity and digital literacy remain significant challenges. Addressing these issues through targeted training programs and infrastructure development is crucial for maximizing AI's potential. Governments, NGOs, and tech companies need to collaborate to bridge these gaps, ensuring that the benefits of AI are accessible to all.

In conclusion, the emergence of AI technologies tailored for small farmers marks a pivotal moment in agriculture. From precise soil management and crop monitoring to advanced irrigation systems and financial tools, AI offers a wealth of solutions that can transform small-scale farming. As these technologies continue to evolve and become more accessible, they hold the promise of ushering in a new era of efficiency, sustainability, and prosperity for small farmers worldwide. The potential for AI to level the playing field in agriculture is immense, and its impact on small-scale farming is a testament to its transformative power.

Empowering Smallholder Farmers

Smallholder farmers are the backbone of agriculture in many parts of the world, often producing a significant portion of the global food supply. But these farmers frequently face a multitude of challenges—limited access to capital, markets, and modern farming techniques. AI presents an unprecedented opportunity to empower smallholder farmers by leveling the playing field and providing them with the tools they need to thrive in an increasingly competitive and resource-constrained environment.

AI-driven technologies can transform the lives of smallholder farmers, enabling them to maximize their resources, reduce waste, and increase productivity. For example, AI can help farmers monitor crop health with precision. Drones equipped with AI capabilities can survey fields in real time, providing farmers with actionable data about soil conditions, moisture levels, and potential disease outbreaks. This technology can replace time-consuming and labor-intensive manual inspections, making it easier for farmers to maintain healthy crops and achieve better yields.

Access to accurate and timely market information is another area where AI can be transformative. Often, smallholder farmers do not have the luxury of real-time data to guide their pricing decisions. AI can analyze vast amounts of information, from local weather patterns to market trends, and offer pricing recommendations that help farmers get the best value for their produce. This not only increases their income but also encourages the efficient use of resources, reducing food waste.

Moreover, AI can provide scalable and customizable educational tools that help farmers learn best practices tailored to their specific needs and conditions. Personalized farming guidance isn't a luxury anymore. With AI-powered mobile apps, farmers can receive tailored advice on everything from planting schedules to pest control, optimized for their unique environmental conditions. These advancements democratize agricultural knowledge, making expert advice accessible to even the most remote and economically disadvantaged farmers.

Financial access is another critical area where AI can make a difference. Many smallholder farmers lack the credit history needed to secure loans. Traditional banking models often overlook them due to high perceived risks and the logistical challenges of remote areas. AI can help bridge this gap by analyzing non-traditional data such as farm productivity, weather patterns, and social metrics to create a more accurate risk profile. This facilitates access to credit and insurance, empowering farmers to invest in better seeds, tools, and technologies.

Collaboration platforms powered by AI can also play a pivotal role. These platforms can match smallholder farmers with buyers, agronomists, and other stakeholders, forming a network that leverages collective knowledge and resources. The result is a more integrated farming community, where resources and information flow more freely, boosting overall productivity and resilience.

AI isn't just about technology; it's about creating ecosystems that support sustainable growth. By improving supply chain logistics, AI can help farmers reduce post-harvest losses, which are a significant challenge for small farms. Enhanced scheduling, route optimization, and predictive demand tools ensure that produce reaches the market at peak freshness, minimizing waste and maximizing profits.

Another promising application of AI is in predictive maintenance of farming equipment. Smallholder farmers may not have the capital to frequently replace machinery. AI can forecast machinery failures before they occur, allowing for timely maintenance and reducing downtimes. This is particularly valuable in peak seasons when every hour of operation counts.

Language and literacy barriers can also be overcome with the help of AI. Voice-activated assistants and chatbots can interact in local languages and dialects, providing instructions and answering questions in a manner that's easy to understand. This ensures that the benefits of AI are accessible to farmers irrespective of their educational background.

It's important to address concerns about data privacy and security. Smallholder farmers need to trust the systems they're using. Transparent data policies and robust cybersecurity measures are crucial for the widespread adoption of AI. Farmers should control their data and know how it's being used, ensuring that their rights are protected while gaining the benefits of AI.

Real-world success stories are numerous, illustrating the transformative potential of AI. In India, AI-powered platforms have helped increase crop yields by providing real-time pest alerts via text messages. In sub-Saharan Africa, AI-based soil sensors are helping farmers optimize fertilizer use, significantly boosting productivity while lowering costs. These examples highlight the practical, on-the-ground benefits that AI can bring to smallholder farmers across varied geographical and socio-economic landscapes.

The role of governments and non-governmental organizations (NGOs) cannot be understated. Policy frameworks that support the adoption of AI in agriculture, combined with initiatives to train and equip farmers, can accelerate the integration of these technologies. Grants, subsidies, and public-private partnerships can lower the financial barriers to entry, making advanced tools accessible to even the smallest farms.

To truly empower smallholder farmers, an inclusive approach is vital. AI developers need to work closely with farming communities to understand their specific needs and challenges. Solutions must be user-friendly, affordable, and practical, tailored to local conditions and cultural contexts. This ensures that AI technologies are not just innovative but also genuinely useful and impactful.

Emerging AI technologies such as machine learning models tailored for low-resource environments, and advanced image recognition for crop diagnosis, are continually evolving. As these technologies become more accessible and easier to use, the gap between smallholder farmers and large-scale agricultural operations will continue to narrow.

Empowering smallholder farmers with AI is about more than just technology; it's about fostering resilience, promoting sustainability, and driving economic growth. It's about giving farmers the tools they need to succeed and ensuring food security for future generations. The journey has just begun, and the possibilities are limitless.

Chapter 19:
Challenges and Limitations

While AI has immense potential to revolutionize agriculture, it's not without its challenges and limitations. Many farmers face significant technological hurdles, particularly in regions where access to high-speed internet and sophisticated hardware is limited. The cost of adopting AI technologies can be prohibitive, especially for small-scale farmers who operate on tight margins. Additionally, there is often a steep learning curve associated with implementing AI-driven solutions, requiring substantial time and resources for training. Beyond practical issues, there is also skepticism and resistance from traditional farming communities who may be wary of over-reliance on technology. Ethical concerns, such as data privacy and the potential for job displacement, further complicate the landscape. Overcoming these challenges will require concerted efforts by governments, industry, and academia to make AI accessible, affordable, and aligned with the needs of diverse agricultural stakeholders.

Technological and Practical Hurdles

Embracing artificial intelligence (AI) in agriculture offers immense potential, but it also comes with a unique set of technological and practical hurdles. One of the standout challenges is the sheer complexity involved in the deployment of advanced AI systems on farms. The agricultural environment is inherently dynamic and unpredictable, making it difficult to develop models that perform reliably under a range of conditions. Weather patterns, soil types, crop varieties, and pest infestations are just a few of the variables that can change quickly, necessitating highly adaptive AI systems.

Data acquisition and management represent another significant barrier. Effective AI systems require vast amounts of high-quality, accurate data, but obtaining this data in agricultural settings can be problematic. Farm environments are often in remote areas where high-speed internet connections are unavailable, leading to difficulties in uploading large datasets for real-time analysis. Moreover, the equipment needed to collect this data, such as drones, sensors, and satellite imagery, can be prohibitively expensive for many farmers, particularly those in developing countries or operating on a smaller scale.

Interoperability between different AI systems and existing farm machinery is also a crucial technological challenge. Most farmers use a variety of tools and technologies, often from different manufacturers. Ensuring that new AI-driven solutions can seamlessly integrate with existing hardware and software is essential for widespread adoption. This requires standardization efforts from manufacturers, which can be slow and fraught with complications. Lack of interoperability can lead to inefficiencies and increased costs, deterring farmers from adopting new technologies altogether.

Furthermore, the robustness of AI algorithms in handling unpredictable farming scenarios is often questioned. Traditional AI models tend to perform well in controlled environments; however, their real-world applicability is limited when confronted with unexpected variables. This discrepancy necessitates ongoing research and field testing, which can be time-consuming and expensive. Equipping AI models to cope with this variability requires sophisticated adaptive learning techniques and continuous updates, presenting a formidable technological hurdle.

On the practical side, the technical skills required to operate and maintain AI technologies are often lacking among farmers and their workforce. While some may be tech-savvy, the majority might not have the requisite expertise to troubleshoot technical issues or optimize system performance. This skills gap necessitates extensive training programs, which can be resource-intensive. Furthermore, older generations of farmers may be resistant to adopting new technologies, creating a cultural barrier that complicates technological implementation.

Another practical consideration is the initial cost of investment in AI technologies. Although these systems promise cost savings and increased efficiencies in the long run, the upfront costs can be daunting. Small to medium-sized farms might find it challenging to allocate the necessary capital, inhibiting the democratization of AI benefits across the agricultural sector. This financial barrier may necessitate government subsidies or innovative financing solutions to make AI technology more accessible to all farmers.

The ethical and regulatory frameworks surrounding AI in agriculture present additional hurdles. The rapid pace of AI development often outstrips existing regulations, leading to a gray area regarding compliance. Farmers and agribusinesses thus find themselves navigating complex legal landscapes, adding another layer of difficulty to the adoption process. This necessitates a collaborative effort between policymakers, technologists, and farmers to develop clear, comprehensive guidelines that can keep pace with technological advancements.

Additionally, data privacy concerns pose a practical challenge. As AI systems collect and analyze vast amounts of data, ensuring that this data is stored securely and used ethically becomes crucial. Farmers may be reluctant to share valuable data about their operations due to fears of misuse or inadequate data protection measures. This skepticism could hinder the collective data-sharing efforts that are essential for refining AI algorithms and generating actionable insights.

Environmental concerns also arise as technological hurdles. AI-powered equipment, such as drones and sensors, often rely on power sources that may not be sustainable. The environmental footprint of manufacturing and deploying these technologies needs thorough examination. Sustainable alternatives must be developed and encouraged to align with the broader goal of promoting environmentally friendly farming practices.

Finally, the disparity in resource availability between developed and developing regions presents both a technological and practical challenge. While farmers in developed countries may have better access to advanced technologies and infrastructure, those in developing regions often face significant constraints. Bridging this gap requires targeted interventions, including international cooperation and investment, to bring the benefits of AI to all farming communities.

Addressing these technological and practical hurdles will require a multi-faceted approach. Governments, researchers, and private enterprises must collaborate to create scalable, adaptable, and affordable AI solutions. As we tackle these challenges, the potential to transform agriculture into an efficient, sustainable, and prosperous sector remains within reach.

Addressing Skepticism and Resistance

As artificial intelligence (AI) continues to permeate the agricultural sector, skepticism and resistance are inevitable. Many farmers and stakeholders grapple with the concept of an algorithm making decisions that were traditionally within human purview. Concerns often stem from a fear of the unknown and perceived threats to traditional farming methods. Convincing these skeptics requires a nuanced approach that addresses their fears and demonstrates AI's tangible benefits.

First, the issue of trust comes to the forefront. Farmers, who often rely on generations of accumulated knowledge and intuitive expertise, may find it difficult to place their faith in technology. Dismissing this skepticism outright would be both unwise and counterproductive. Instead, highlighting case studies where AI has led to significant positive outcomes can be highly effective. For instance, sharing stories of farmers who have successfully integrated AI to increase crop yields or reduce resource wastage can make the benefits more relatable and tangible.

Another significant factor contributing to resistance is the fear of job displacement. Many farmers and agricultural workers worry that AI and automation will render their roles obsolete. Addressing this fear involves emphasizing that AI is designed to complement human labor rather than replace it. AI can take over repetitive and strenuous tasks, freeing up human workers for more complex, decision-making roles. This can lead to enhanced job satisfaction and a safer working environment.

However, integrating AI in agriculture isn't without challenges. The initial costs of adopting AI technologies can be prohibitive for small-scale farmers. Many are hesitant to make such significant investments without guaranteed returns. Providing evidence of long-term cost savings and efficiency improvements is crucial in alleviating these concerns. Moreover, creating financial models or subsidies that support initial AI adoption can go a long way in encouraging uptake.

Moreover, addressing data privacy and ethical concerns is vital. Farmers may be apprehensive about sharing their data with AI platforms, fearing misuse or unauthorized access. Ensuring that AI solutions adhere to stringent data protection standards and providing clear, transparent information about how data will be used can help build trust. Furthermore, implementing robust safeguards to prevent data breaches will reassure users that their information is secure.

Training and education are equally important. Many in the agricultural sector might resist AI simply due to a lack of understanding about how it works and its potential benefits. Offering training programs, workshops, and educational resources aimed at demystifying AI can empower farmers to make informed decisions. Collaborative efforts with agricultural extension services, universities, and tech companies to provide hands-on training can bridge knowledge gaps and reduce resistance.

Resistance can also be rooted in cultural factors. Farming is often a deeply ingrained way of life, with practices passed down through generations. For some, adopting AI could feel like abandoning a part of their heritage. Engaging with these farmers empathetically and respectfully is essential. Showing how AI can enhance traditional methods rather than replace them can help change perceptions. For example, precision agriculture technologies can optimize the use of age-old farming techniques, making them more sustainable and efficient.

It's also crucial to acknowledge that AI will not be a one-size-fits-all solution. Different regions and types of farms have unique challenges and needs. Flexibility in AI solutions allows for customization to fit specific contexts. This adaptability can alleviate concerns that AI might be too rigid or unsuitable for certain types of farming.

While technical hurdles exist, they are not insurmountable. Connectivity issues in rural areas, for instance, can hinder the adoption of AI technologies that require real-time data access. This is slowly being mitigated by advancements in rural internet infrastructure and the development of AI solutions that can function offline or with intermittent connectivity. Demonstrating that these technical challenges are being actively addressed can help ease resistance.

Lastly, fostering a community around AI in agriculture can help in overcoming skepticism. Creating forums where farmers can share their experiences, challenges, and successes with AI can build a support network. Peer validation often plays a crucial role in technology adoption, as farmers are more likely to trust and follow the lead of their peers.

In summary, overcoming skepticism and resistance to AI in agriculture requires a multifaceted approach. Building trust through transparent practices, offering financial and educational support, respecting cultural values, and addressing technical challenges are all crucial steps. By focusing on these areas, we can pave the way for AI to be seen not as a threat, but as a powerful ally in creating a more sustainable and productive agricultural future.

Chapter 20:
Future Trends in AI Agriculture

The next frontier in AI agriculture promises a realm of innovations that could redefine how we cultivate and manage our food resources. Emerging technologies like edge computing, blockchain integration, and advanced robotics are set to transform farming into a hyper-efficient, data-driven industry. These trends will enable precise crop management, automate intricate farming processes, and optimize resource utilization. Moreover, leveraging AI for climate resilience will become critical as climate change poses new challenges. Farmers can look forward to sophisticated decision support systems that predict market trends, pest infestations, and weather patterns with unparalleled accuracy. Embracing these advancements, the future of farming holds immense potential to boost productivity sustainably and create a resilient agricultural ecosystem.

Emerging Technologies and Innovations

In the rapidly evolving landscape of AI-driven agriculture, emerging technologies stand at the forefront, bringing unprecedented innovations and solutions to age-old farming challenges. These advancements are creating new pathways for enhancing productivity, sustainability, and efficiency in agricultural practices. One of the most groundbreaking innovations is the integration of IoT (Internet of Things) with AI systems. By embedding sensors in fields, IoT devices can capture real-time data on soil health, plant growth, and weather conditions. This data is then analyzed by AI algorithms to provide actionable insights, ensuring farmers can make data-driven decisions to optimize yields.

Another transformative technology is the advent of advanced robotics, powered by sophisticated AI algorithms. These robots are not limited to just performing repetitive tasks; they're now capable of complex behaviors such as identifying ripe fruits, detecting nutrient deficiencies in plants, and even autonomously planting seeds. Companies are developing robots that can navigate through fields with pinpoint accuracy, reducing the need for human intervention. These advancements mark a significant shift from traditional manual labor to highly automated, precise farming techniques.

Biotechnology combined with AI is opening new frontiers in crop improvement. Machine learning models are now being utilized to analyze genetic data, allowing scientists to identify traits that lead to more resilient and high-yield crops. AI-driven gene editing technologies, such as CRISPR, are enabling precise modifications in the plant genome, fostering the development of crops that can withstand diseases, pests, and changing climate conditions. This synergy between AI and biotech offers immense potential for addressing global food security challenges.

The potential of blockchain technology in agriculture is also gaining traction. When combined with AI, blockchain can provide unparalleled transparency and traceability in the supply chain. Smart contracts, powered by AI, can automate and authenticate transactions between different stakeholders, ensuring fair trade practices. Additionally, blockchain can securely record the entire lifecycle of a crop—from planting to consumption—giving consumers confidence in the authenticity and quality of their food.

In addition to these innovations, the development of sophisticated AI software tools tailored for agricultural use is transforming farm management practices. These tools integrate various data points—from satellite imagery to weather forecasts—to offer comprehensive farm management solutions. Farmers can now monitor crop growth, predict harvest times, and even estimate market prices, all through user-friendly interfaces. This democratization of technology ensures that even small-scale farmers can harness the power of AI to improve their farming operations.

Moreover, the advent of digital twin technology in agriculture is another revolutionary development. A digital twin is a virtual replica of a physical entity, in this case, a farm. By creating digital twins, AI can simulate various scenarios, such as predicting the impact of weather changes or pest infestations. These simulations provide farmers with valuable foresight, allowing them to take preventive measures rather than reactive ones.

Machine vision technology is also playing a crucial role in advancing AI agriculture. With the help of high-resolution cameras and advanced image processing algorithms, AI systems can now monitor crop health in real time. This technology can detect early signs of disease, nutrient deficiencies, and pest infestations, enabling timely interventions. Machine vision is not just limited to crops; it's also being used in livestock management to monitor the health and behavior of animals.

Furthermore, the integration of AI with cloud computing is empowering farmers with real-time data access and analysis. Cloud-based platforms offer scalable solutions, enabling farmers to store and analyze large volumes of data without the need for significant local computing resources. These platforms also facilitate collaboration among different stakeholders, from agronomists to supply chain managers, fostering a more connected and efficient agricultural ecosystem.

The development of specialized drones equipped with AI capabilities is yet another innovation reshaping the agricultural landscape. These drones can perform a variety of tasks, such as aerial spraying, crop monitoring, and even planting seeds. By leveraging AI, drones can analyze the data they collect to provide detailed insights into crop health and field conditions. This high level of precision and efficiency reduces the need for manual labor and ensures optimal resource utilization.

In the realm of environmental sustainability, AI is driving innovations that promote eco-friendly farming practices. Precision irrigation systems, guided by AI, are optimizing water usage, thereby conserving this valuable resource. AI algorithms analyze soil moisture data and weather patterns to determine the exact amount of water required, minimizing waste. Similarly, AI is being used to develop organic farming methodologies that reduce the reliance on chemical pesticides and fertilizers, promoting soil health and biodiversity.

Another fascinating development is the use of AI in vertical farming, where crops are grown in stacked layers within controlled environments. AI systems monitor and regulate factors such as light, temperature, and humidity, ensuring optimal growth conditions. This innovative approach not only maximizes space utilization but also allows for year-round cultivation, independent of weather conditions. Vertical farming, powered by AI, holds the promise of urban agriculture, bringing fresh produce closer to city dwellers.

The combination of AI and augmented reality (AR) is also beginning to make waves in the agricultural sector. Through AR-enabled devices, farmers can visualize data overlaid on their physical environment. For instance, an AR headset might display information about plant health directly on the plants themselves as a farmer walks through the field. This immersive technology makes data more accessible and actionable, enhancing decision-making processes.

Collaborations between AI startups and traditional agricultural companies are accelerating the pace of innovation. These partnerships bring together technological expertise and sector-specific knowledge, fostering the development of tailored solutions that address real-world farming challenges. The cross-pollination of ideas and resources in these collaborations is critical for scaling AI innovations and making them accessible to farmers globally.

As AI continues to evolve, the future of agriculture looks increasingly promising. Innovations are not only aimed at improving productivity but also at making farming more sustainable and resilient to climate change. From the lab to the field, AI is unlocking new possibilities, transforming agriculture into a high-tech industry capable of meeting the challenges of the 21st century. The journey of integrating AI into agriculture is still in its nascent stages, but the pace at which technology is advancing suggests a future where intelligent systems will be an integral part of every farm.

In summary, the emerging technologies and innovations in AI agriculture are setting the stage for a new era of smart farming. By leveraging the power of IoT, robotics, biotechnology, blockchain, machine vision, cloud computing, drones, and more, AI is revolutionizing the way we grow food. These advancements are not just enhancing efficiency and productivity but are also paving the way for sustainable and resilient agricultural practices. As we prepare for the future of farming, the integration of these cutting-edge technologies will undoubtedly play a critical role in driving the next green revolution.

Preparing for the Future of Farming

As the agricultural sector stands on the brink of a technological transformation, the significance of preparing for the future of farming cannot be overstated. Artificial Intelligence (AI) promises to revolutionize the way we cultivate crops, manage livestock, and harness natural resources. This evolution isn't just about adopting cutting-edge technologies; it's about rethinking traditional practices and aligning them with a future that's increasingly data-driven and automated. The continued success of farming is contingent upon readiness and adaptability in the face of these rapid advances.

One of the first steps in preparing for the future is to cultivate a mindset that embraces change. Farmers, technologists, and policymakers must work together to foster an environment where innovation thrives. This requires an openness to learning and a willingness to experiment with new technologies, even if they challenge long-standing practices. By fostering a culture of continuous education and training, individuals in the agricultural sector can stay abreast of technological advancements and their practical applications.

Investing in AI infrastructure is another critical aspect. From AI-powered drones that monitor crop health to autonomous machinery that performs planting and harvesting, the integration of AI tools demands robust technological infrastructure. This includes high-speed internet for real-time data transmission, cloud storage solutions for data management, and advanced sensors and devices for precise monitoring. Ensuring that farms are equipped with these essentials will lay the groundwork for the seamless adoption of AI technologies.

Equally important is the establishment of comprehensive data management systems. The future of farming will rely heavily on big data to make informed decisions. Both small and large farms must implement strategies for collecting, storing, and analyzing data. This isn't merely about acquiring data but ensuring its quality and relevance. Farmers need to understand which data points are crucial – whether it's soil moisture levels, weather patterns, or pest infestations – and how to leverage this information to optimize yields and sustainability.

Collaboration plays a pivotal role in preparing for the future. Partnerships between farmers, technology companies, academic institutions, and government bodies can accelerate the development and deployment of AI solutions. Collaborative efforts can lead to the sharing of resources, knowledge, and expertise, ultimately fostering innovation. For instance, research universities can work with tech companies to tailor AI solutions to specific agricultural needs, while government policies can provide the necessary support and incentives.

In addition to technological and collaborative readiness, there's a pressing need to address regulatory and ethical considerations. As AI systems become more entrenched in farming practices, questions around data privacy, security, and ethical usage will arise. Establishing clear guidelines and standards will ensure that AI applications are both effective and ethically sound. This involves crafting regulations that protect farmers' data while allowing for innovation and competition among tech providers.

Furthermore, the development of AI literacy among farmers is essential. Educational programs and workshops focusing on AI applications in agriculture will empower farmers with the knowledge and skills needed to leverage these technologies effectively. Training programs should cover a breadth of topics, from understanding basic AI concepts to hands-on sessions with AI tools and software. By demystifying AI and making it accessible, we can mitigate resistance and uncertainty, paving the way for smoother adoption.

Preparing for the future also means embracing sustainability and resilience. AI technologies can contribute to sustainable farming practices by optimizing resource use, reducing waste, and monitoring environmental impact. For instance, AI-driven irrigation systems can pinpoint precisely when and where water is needed, conserving this precious resource. Similarly, predictive analytics can help anticipate and mitigate adverse environmental impacts, making farms more resilient to climate change and other challenges.

The potential economic benefits of AI in agriculture are immense, but realizing these gains requires strategic planning and investment. Farmers must evaluate the costs and benefits of AI technologies, considering both the short-term expenses and the long-term gains in efficiency and productivity. Financing options, such as government grants, subsidies, and private investment, can ease the financial burden, making it feasible for farms to adopt these new technologies.

Lastly, fostering a supportive community is vital. Farmers should engage in networks and forums where they can share experiences, challenges, and successes with AI adoption. Peer support can be incredibly powerful, as it provides real-world insights and fosters a sense of community among those navigating similar challenges. By building a cohesive support system, farmers can collectively drive the advancement and acceptance of AI in agriculture.

The future of farming hinges not just on the technologies themselves but on our preparedness to integrate them seamlessly into our agricultural landscapes. With the right mindset, infrastructure, education, and collaboration, we can harness the power of AI to foster a more sustainable, efficient, and productive farming industry. While the challenges are significant, the potential rewards far outweigh the effort required to achieve this technological evolution.

As we envision the farms of tomorrow, we must remain grounded in the reality of today's challenges and opportunities. The transition to AI-powered agriculture won't be instantaneous but through deliberate planning and collective effort, we can shape a future where technology and tradition coexist harmoniously for the greater good of our global food systems.

Chapter 21:
Educating the Next Generation

The future of agriculture lies in the hands of the next generation, and equipping them with the right knowledge and tools is paramount for sustained progress. Schools and universities must integrate AI-focused agricultural curriculums, covering everything from machine learning basics to advanced AI applications. Training programs and workshops tailored for young farmers can bridge the gap between traditional practices and modern technologies, fostering an environment of innovation and adaptability. By emphasizing hands-on learning and real-world problem-solving, we can inspire a new wave of agri-tech pioneers who are not only skilled in AI but also deeply passionate about sustainable farming practices. This commitment to education ensures that the collective wisdom of today's advancements propels the agricultural sector into a more efficient, productive, and environmentally conscious future.

AI Curriculum for Agriculture

As we venture into educating the next generation about artificial intelligence in agriculture, it's crucial to establish a comprehensive curriculum that bridges technology with traditional farming practices. This curriculum aims to equip aspiring farmers, data scientists, and agronomists with the knowledge and skills needed to leverage AI for more sustainable and efficient agricultural practices. By combining the fundamentals of AI with real-world applications in agriculture, we can cultivate a new breed of agricultural professionals who are not only tech-savvy but also deeply committed to sustainable farming.

First and foremost, the curriculum should introduce the **Basics of AI and Machine Learning**. Students need a strong foundation in understanding what AI is, the different types of machine learning, and how these technologies can be utilized to process vast amounts of agricultural data. The aim here is to demystify AI and make it accessible, focusing on practical examples that relate directly to the agricultural sector. This foundational knowledge will be essential as they move on to more advanced topics.

Once the basics are covered, the curriculum should shift its focus to *Applied AI in Agriculture*. This includes detailed modules on precision agriculture, which is all about making farming more controlled and accurate thanks to big data and satellite technology. We can go further by introducing students to real-time data collection and analysis techniques, teaching them how to gather and interpret data from sensors, drones, and satellite images. These skills are invaluable for making informed decisions that enhance crop yields and reduce resource wastage.

Next, students should explore the specific AI applications for **Crop Monitoring and Management**. Here, they'll learn how AI can identify diseases in crops early on, assess crop health, and even predict harvest times. Through hands-on projects, students can work with AI models to diagnose plant health issues, using image recognition technologies to spot signs of blight, mildew, or pest infestations before they spread. These practical skills are not only engaging but also directly applicable to managing a healthy and productive farm.

Soil health is another critical area where AI can make a significant impact. In the **Soil Analysis and Health** module, students will learn about AI-driven soil testing and how to interpret soil data to optimize conditions for various crops. This section will cover the essentials of soil chemistry, texture, and moisture levels, along with how AI algorithms can recommend specific fertilizers and soil amendments to enhance soil productivity.

Water management is yet another pivotal aspect, addressed in the **Water Management** module. Effective irrigation strategies are the backbone of successful farming operations, especially in areas prone to droughts. Students will be introduced to AI solutions for monitoring water usage and maintaining optimal soil moisture levels. By learning about automated irrigation systems paired with AI, they'll be better equipped to manage water resources efficiently, ensuring crops receive the right amount of water at the right time.

Controlling weeds and pests is a continuous challenge in agriculture, one that AI is uniquely positioned to address. In the **Weed and Pest Control** module, students will explore AI techniques for identifying and managing weeds, as well as pest prediction and management strategies. With detailed case studies and practical exercises, they'll see how AI can identify pest patterns and suggest timely interventions, thus minimizing crop damage and reducing the need for chemical pesticides.

The curriculum wouldn't be complete without covering the significant advancements in **Autonomous Machinery**. With AI-powered tractors, harvesters, and planting robots, the face of modern farming is rapidly changing. Students will delve into the workings of these autonomous machines and understand how machine learning algorithms enable them to make decisions in real-time. Through interactive simulations and hands-on activities, they'll gain insights into operating and maintaining this advanced machinery.

Moving on, the curriculum should also include a section on *Predictive Maintenance* for agricultural equipment. Here, students will learn how AI can predict equipment failures before they occur, thus reducing downtime and maintenance costs. This involves understanding the various sensors and data analytics tools that monitor the health of machinery and predict when maintenance is required.

Students will also need a grasp of how AI can optimize the **Supply Chain**. In the *Supply Chain Optimization* module, they'll explore how AI helps in enhancing logistics, improving storage, and ensuring timely distribution of agricultural products. This section will highlight the benefits of transparent and efficient supply chains, reducing food waste, and meeting market demands effectively.

To make informed choices, farmers need accurate market predictions. Hence, the curriculum should cover **Market Analysis and Decision Making**. Students will learn how AI can analyze market trends, predict prices, and assist in making data-driven decisions about what crops to plant and when to harvest. They'll explore various AI tools and software that provide market insights, ensuring that they understand both the technological and economic aspects of farming.

Incorporating AI in *Smart Greenhouses* is another intriguing area of study. This part of the curriculum will delve into how AI can create controlled environments that maximize crop yields while minimizing resource use. Students will learn about sensors, climate control systems, and AI algorithms that adjust conditions in real-time to create optimal growing conditions.

Understanding AI's role in **Livestock Management** is crucial for those involved in animal husbandry. The curriculum should include modules on using AI for tracking animal health, productivity, and efficient feeding systems. Students will gain practical knowledge on how AI can monitor livestock, predict health issues, and automate feeding schedules to ensure optimal animal welfare and productivity.

A significant portion of the curriculum should be dedicated to the **Economic and Environmental Benefits** of AI in agriculture.

Students will explore how AI can reduce costs, increase efficiency, and promote environmental sustainability. They'll be introduced to concepts like carbon footprint reduction, sustainable farming practices, and how AI can contribute to promoting biodiversity and conservation efforts.

As future leaders in the agricultural sector, students must be aware of the **Policy and Regulation** landscape. This part of the curriculum will cover how to navigate agricultural policies related to AI, ethical considerations, and data privacy concerns. By understanding the regulatory environment, they can better advocate for policies that support sustainable and innovative farming practices.

Finally, to tie all these elements together, the program should encourage students to engage with **Global Case Studies**. By examining successful implementations of AI in diverse climates and cultures, students can gain valuable insights and lessons that can be applied to their contexts. This global perspective fosters an understanding that, while technology can be universally beneficial, its application may need to be tailored to meet specific regional challenges and opportunities.

As AI continues to reshape the agricultural landscape, educating the next generation with a robust, forward-thinking curriculum is essential. By providing a blend of theoretical knowledge and practical skills, this AI curriculum for agriculture aims to create innovators who are ready to tackle the challenges of modern farming and lead the next green revolution.

Training Programs and Workshops

The transformative power of AI in agriculture is only as strong as the people who understand and implement it. To truly harness this power, it is essential to focus on comprehensive training programs and workshops that equip farmers, agronomists, and agricultural technicians with the skills and knowledge they need. These educational initiatives not only demystify AI but also inspire innovation and a proactive approach to integrating new technologies.

Training programs tailored for various expertise levels can help bridge the gap between the potential of AI and its practical application in the field. For beginners, introductory courses could cover the basics of AI and its relevance to agriculture. These programs can offer hands-on sessions where participants learn to operate AI-powered tools and interpret the data these tools generate. Workshops can also spotlight case studies showcasing successful implementations of AI, setting a roadmap for what's possible.

One effective approach is to create immersive learning environments where participants can engage directly with AI technologies. Smart farms, equipped with the latest AI-driven tools and machinery, can serve as live classrooms. Here, trainees can observe how AI optimizes various farming processes, from soil analysis to crop monitoring. This experiential learning can be incredibly beneficial, turning theoretical knowledge into practical skills.

Moreover, the curriculum should be flexible enough to accommodate the rapidly evolving nature of AI technology. Continuing education opportunities, such as advanced workshops and certification programs, ensure that even seasoned professionals remain up-to-date with the latest advancements. These programs can delve deeper into specialized areas like autonomous machinery and predictive maintenance, helping participants develop niche expertise.

Collaborating with agricultural universities and tech institutions is another promising strategy. By integrating AI-related modules into agricultural science programs, we can nurture a new generation of farmers and agronomists who are AI-literate from the outset. Such collaborations can also lead to joint research projects and the development of innovative solutions tailored to specific agricultural challenges.

Interactive online platforms can extend the reach of these training programs beyond geographic limitations. E-learning modules, webinars, and virtual workshops can provide flexible learning options for individuals who may not have the time or resources to attend in-person sessions. These platforms can also foster a global community of like-minded individuals, allowing for the exchange of ideas and best practices across borders.

Peer-to-peer learning can be another powerful tool. Farmers who have successfully integrated AI into their operations can share their experiences and insights through workshops or mentorship programs. This approach not only builds a sense of community but also encourages a culture of innovation and continuous improvement.

Furthermore, industry-sponsored workshops can bring together experts from various fields, offering a multidisciplinary perspective on AI in agriculture. These workshops can feature speakers from tech companies, academic researchers, and successful farmers, providing a holistic view of the AI landscape. Sponsorship can also help cover costs, making these valuable learning opportunities more accessible to a broader audience.

Government and non-governmental organizations (NGOs) can play a crucial role in facilitating these training programs. By providing funding, resources, and logistical support, they can help scale these initiatives, particularly in regions where access to such educational resources is limited. Policy frameworks that encourage lifelong learning and skills development can further support these efforts, ensuring that training programs remain sustainable and impactful.

Practical demonstrations and participatory workshops can also be highly effective. Events such as field days and demonstration farms allow participants to see AI technologies in action. These hands-on experiences can demystify complex concepts and show the tangible benefits of AI, making the technology more approachable and easier to adopt.

Feedback mechanisms should be an integral part of any training program. Collecting feedback from participants can provide valuable insights into the effectiveness of the training and highlight areas for improvement. This iterative process ensures that the programs remain relevant and engaging, continually evolving to meet the needs of the agricultural community.

At the end of the day, the goal of these training programs and workshops is not only to impart technical skills but also to inspire a mindset shift. Embracing AI in agriculture requires a willingness to innovate and adapt. By providing comprehensive education and continuous support, we can empower the next generation to lead the way in a new era of sustainable and efficient farming.

As AI continues to redefine the agricultural landscape, these educational initiatives will be the cornerstone of this transformation. Investing in training programs and workshops now will yield long-term benefits, fostering a workforce that is not only skilled but also enthusiastic about the possibilities AI brings to agriculture.

Chapter 22:
Collaborations and Partnerships

The transformative impact of AI in agriculture hinges significantly on effective collaborations and partnerships. Governments and NGOs play a pivotal role in creating a conducive environment through supportive policies and funding initiatives, facilitating the integration of AI technologies in farming practices. Concurrently, industry-academia partnerships foster innovation by combining cutting-edge research with practical agricultural applications. These alliances promote knowledge sharing, drive technological advancements, and ensure that AI solutions are both practical and scalable. The synergy between various stakeholders accelerates the adoption of AI, fostering a networked ecosystem where resources are efficiently utilized, and innovations are rapidly translated from lab to field. As these collaborations deepen, they hold the promise of unlocking unprecedented productivity and sustainability in agriculture, fostering resilience in the face of global challenges.

Role of Governments and NGOs

The intersection of artificial intelligence (AI) and agriculture is a dynamic and rapidly evolving field. As technology becomes more sophisticated, its potential to revolutionize farming practices grows exponentially. However, the full realization of AI's promise in agriculture is not solely dependent on technological advancements. The role of governments and non-governmental organizations

(NGOs) is equally critical in fostering an environment where these innovations can truly thrive.

Governments play a pivotal role in shaping the framework within which AI technologies can be adopted in agriculture. From policy development to funding research initiatives, their influence extends across multiple dimensions. Policies that encourage the integration of AI into agricultural practices can accelerate innovation and ensure equitable access to these advanced tools, particularly for small-scale farmers. For instance, subsidies or tax incentives for adopting precision agriculture technologies can make these tools more accessible.

NGOs, on the other hand, serve as crucial intermediaries that bridge the gap between technology developers and the farming community. They often focus on educational initiatives aimed at familiarizing farmers with new AI technologies. By conducting workshops, training sessions, and pilot programs, NGOs can help demystify AI and demonstrate its practical applications on the ground. This educational outreach is vital for overcoming skepticism and resistance to change, which are common hurdles in the adoption of new technologies.

The collaboration between governments and NGOs is essential for the holistic development of AI in agriculture. Governments can provide the regulatory and financial backbone, while NGOs can offer the grassroots support needed to implement these technologies effectively. This symbiotic relationship ensures that policies are not just theoretical but are practically applied and tested in various farming contexts. Such collaborations can also facilitate data sharing and the creation of centralized databases, which are crucial for the effective functioning of AI algorithms in agriculture.

Moreover, governments and NGOs can jointly fund research and development projects that tackle specific agricultural challenges using AI. These projects can range from developing AI models for predicting crop diseases to creating automated systems for precision irrigation. By pooling resources and expertise, they can spur innovations that might be too risky or expensive for private entities to undertake independently. This collaborative R&D can lead to breakthroughs that benefit the entire agricultural sector.

Another critical area where government and NGO collaboration can make a significant impact is in the realm of policy advocacy. Advocacy efforts can highlight the need for robust data privacy laws that protect farmers' interests while enabling the seamless integration of AI technologies. Policies that address data ownership, use, and sharing are crucial for building trust among farmers, who might be wary of how their data is being used by technology companies.

Effective collaboration also extends to addressing the digital divide that exists in many rural areas. Governments and NGOs can work together to improve digital infrastructure, ensuring that farmers have access to the internet and digital tools necessary for AI technologies. This includes not just physical infrastructure like broadband but also digital literacy programs that equip farmers with the skills they need to leverage these tools effectively.

It's also worth noting the role of international collaborations. Agricultural challenges are often global in nature, such as climate change, pest invasions, and food security. Governments and NGOs can participate in international consortia that share knowledge, technologies, and best practices. Such global partnerships can facilitate the transfer of AI technologies to regions that need them the most, thereby promoting global agricultural resilience.

The role of governments and NGOs extends to fostering a culture of innovation and sustainability. By promoting practices that are both technologically advanced and environmentally sustainable, they can help ensure that AI technologies contribute to long-term agricultural productivity without depleting natural resources. Sustainability grants and awards can incentivize the development and adoption of AI solutions that address environmental concerns, such as reducing pesticide use or minimizing water waste.

Furthermore, NGOs often play a critical role in advocacy and policy reform. They can work with governments to propose regulations that ensure fair competition within the agri-tech industry, protect smallholder farmers, and promote the ethical use of AI. NGOs can also act as watchdogs, holding both public and private entities accountable for their impacts on the agricultural sector.

In conclusion, the role of governments and NGOs is indispensable in the transformative journey of AI in agriculture. Their collaborative efforts can pave the way for a more innovative, equitable, and sustainable agricultural landscape. Through policy support, educational initiatives, and global partnerships, they can help unlock the full potential of AI, driving the next green revolution and ensuring that its benefits are widely shared.

Industry-Academia Partnerships

The synergy between industry and academia is reshaping the landscape of modern agriculture, driving innovation at an unprecedented pace. Industry-academia partnerships are pivotal in ensuring that the latest advancements in artificial intelligence (AI) are seamlessly integrated into agricultural practices. These collaborations serve as a bridge, facilitating the transfer of cutting-edge technologies from research labs to farms, thereby making sustainable and efficient farming practices more accessible for farmers.

Universities and research institutions play an essential role in these partnerships. They are the breeding grounds for new ideas and technological solutions. Academia contributes not only through groundbreaking research but also by developing the essential theoretical frameworks that underpin innovative AI applications in agriculture. Researchers and students engaged in projects related to machine learning, robotics, and data analytics frequently collaborate with industry experts to create practical solutions tailored for agricultural needs.

One illustrative example is the development of AI-driven disease prediction models. Researchers rely on large datasets to train these models, requiring close collaboration with agricultural firms that can provide the necessary data. These partnerships allow for real-time testing and refinement of technologies, ensuring they are ready for practical application. Farmers, in turn, benefit from these cutting-edge tools that predict crop diseases, enabling them to take preventive measures and minimize crop loss.

While academia brings scientific rigor and innovation, industry partners offer the resources and real-world testing environments that academic institutions often lack. Companies have the infrastructure and funding to scale innovations from pilot projects to widespread implementation. This symbiotic relationship accelerates the development cycle of new technologies, making it possible to see a tangible impact on the field much sooner.

Moreover, collaborations between industry and academia are not confined to just the development of new technologies. These partnerships extend to educational initiatives aimed at preparing the next generation of farmers, researchers, and technologists. Universities, with support from industry partners, are increasingly offering specialized courses and training programs focused on AI in agriculture. These programs equip students with both theoretical knowledge and practical skills, ensuring they are ready to tackle the complex challenges of modern farming.

For instance, several universities have initiated AI-focused agricultural programs where students work on real-world projects provided by industry partners. These collaborative projects serve as capstones, allowing students to apply their learning directly to industry-relevant problems. This hands-on approach not only enhances the students' learning experience but also provides the industry with fresh perspectives and innovative solutions.

Additionally, funding from industrial partners is often vital in advancing academic research. Many universities rely on grants and sponsorships from agricultural companies to sustain their research activities. This financial support allows academic institutions to undertake ambitious projects that require substantial resources, such as the development of autonomous machinery or advanced soil health monitoring systems. In return, companies gain early access to groundbreaking discoveries and technologies that can give them a competitive edge in the agricultural market.

Industry-academia partnerships also play a critical role in addressing global agricultural challenges. Collaborative research often focuses on pressing issues like climate change, food security, and sustainable farming practices. By pooling their expertise and resources, these partnerships can develop holistic solutions that address multiple facets of these complex problems. For instance, a collaborative project might combine climate data analysis, crop modeling, and AI algorithms to create an integrated system for optimizing crop yields in the face of changing environmental conditions.

Furthermore, these partnerships often extend beyond national borders, involving multiple institutions from different countries. Global collaborations bring together diverse expertise and perspectives, fostering innovation that is both inclusive and comprehensive. For example, multinational projects on water management can leverage the nuanced understanding of local conditions provided by academic institutions from different regions, resulting in more adaptable and effective AI solutions.

The collaborative efforts of industry and academia are also crucial in overcoming the practical challenges of implementing AI technologies in agriculture. Farmers often face barriers such as high costs, technological complexity, and lack of familiarity with new tools. Through joint initiatives, industry and academia can develop user-friendly, cost-effective solutions that are designed with the end-user in mind. Demonstration farms, pilot projects, and extension programs are some of the ways in which these partnerships can facilitate technology adoption among farmers.

It's also important to note the role of interdisciplinary collaboration in these partnerships. Agricultural innovation increasingly relies on the integration of knowledge from different fields, including computer science, engineering, biology, and environmental science. Industry-academia collaborations often bring together experts from these diverse disciplines, fostering a multidisciplinary approach to problem-solving. This holistic perspective is essential for developing robust AI solutions that are practical and sustainable.

To illustrate, a project aimed at developing AI-powered irrigation systems might involve hydrologists, computer scientists, agronomists, and engineers working together. Each expert brings a unique set of skills and knowledge, contributing to a comprehensive solution that optimizes water usage while maintaining soil health and crop productivity. These interdisciplinary partnerships ensure that technological innovations are scientifically sound, practically viable, and environmentally sustainable.

The future of industry-academia partnerships in agriculture looks promising, with many exciting developments on the horizon. As AI technologies continue to evolve, these collaborations will be crucial in ensuring that advances translate into impactful solutions on the ground. From innovative pest management systems to precision agriculture tools, the combined efforts of industry and academia will drive the next wave of agricultural innovation, paving the way for a more sustainable and efficient farming future.

In conclusion, industry-academia partnerships are a cornerstone of agricultural innovation, bridging the gap between research and practical application. By leveraging the strengths of both sides, these collaborations accelerate the development and adoption of AI technologies in agriculture. They not only drive technological advancements but also ensure their effective implementation, ultimately contributing to the transformation of modern farming. As we look towards the future, the continued collaboration between industry and academia will be essential in unlocking the full potential of AI in agriculture, leading us towards a new era of sustainable and intelligent farming practices.

Chapter 23:
Funding and Investment

Securing funding for AI-driven agricultural projects is a crucial step in accelerating the adoption of innovative technologies that promise to revolutionize farming practices. Investors and venture capitalists are increasingly recognizing the potential of AgriTech to transform the agricultural sector, making it more efficient, sustainable, and productive. From government grants and subsidies aimed at fostering technological advancements in farming to private equity investments focusing on scalable AgriTech startups, there are multiple avenues to explore. Crowdfunding platforms are also emerging as viable options for smaller projects, providing opportunities for tech-savvy individuals and communities to participate in this green revolution. As the interest in sustainable agriculture grows, so does the pool of potential investment, paving the way for groundbreaking innovations that can tackle pressing challenges such as food security, climate change, and resource management.

Securing Funding for AI Projects

Securing funding for AI projects in agriculture is a critical step for innovators aiming to transform traditional farming practices. Given the substantial initial investment needed for research, development, and deployment of AI technologies, it's essential to navigate the funding landscape effectively. This section will explore the various avenues available to secure financial backing and practical strategies to convince potential investors of AI's transformative impact on agriculture.

Agritech startups can consider a mix of funding sources—government grants, venture capital, angel investors, and corporate partnerships. Government grants play a pivotal role, especially in regions committed to revamping their agricultural sectors through technology. Numerous grant programs are available at the national and regional levels, targeting projects that promise sustainability, increased productivity, and environmental benefits.

Pursuing venture capital is another viable route. Venture capital firms are increasingly showing interest in agritech startups, recognizing the immense potential for growth and innovation in this sector. When pitching to venture capitalists, it's essential to highlight your AI project's scalability, sustainability, and the tangible benefits it brings to the farming community. Demonstrating a clear path to profitability is equally crucial, as investors need assurance that their capital will yield significant returns.

Angel investors, typically wealthy individuals with a keen interest in supporting startups, can also be instrumental. They often offer more than just financial backing; their mentorship and industry connections can be invaluable. The key to attracting angel investors is to present a compelling vision and a robust business plan. In many cases, angel investors are drawn to projects that align with their personal values, such as reducing environmental impact or enhancing food security.

Corporate partnerships offer another funding avenue. Many companies in the agricultural sector are eager to embrace AI solutions that can enhance their efficiency and profitability. By partnering with established corporations, startups can benefit from financial support, market access, and industry expertise. It's a symbiotic relationship;

corporations gain cutting-edge technologies, while startups secure the resources needed to scale their operations.

To maximize funding opportunities, it's essential to build a strong case for your AI project. This involves demonstrating the technology's efficacy through pilot projects or prototypes. Real-world evidence of AI applications in agriculture, such as improved crop yields, efficient water management, or enhanced pest control, can be persuasive. Data-driven proof of concept can significantly bolster your pitch to investors.

Moreover, communicating the broader impact of your AI project can resonate with potential funders. Highlighting how your technology addresses pressing global challenges, such as food security and climate change, can make your proposal more attractive. Investors are increasingly looking for projects that offer not just financial returns but also social and environmental benefits.

Another critical aspect of securing funding is networking and building relationships within the agricultural and tech communities. Attending industry conferences, participating in pitch competitions, and joining relevant networking groups can open doors to funding opportunities. Establishing a presence in these circles helps you connect with potential investors and partners who share your vision.

It's also important to stay informed about new funding trends and emerging opportunities. Crowdfunding platforms have become popular for raising small amounts of money from a large number of people. While this may not suffice for large-scale AI projects, it can provide initial funding and validate your idea in the market.

Additionally, incubators and accelerators focused on agritech can be valuable resources. These programs often provide seed funding, mentorship, and access to networks that can help your project grow.

Being part of an accelerator also lends credibility to your startup, making it more attractive to future investors.

Pursuing funding is a competitive process. Tailoring your pitch to the specific interests and criteria of each potential funder increases your chances of success. Customizing your proposals, maintaining transparency about risks and potential returns, and presenting a clear implementation timeline are strategies that can help secure the financial backing you need.

In conclusion, securing funding for AI projects in agriculture requires a strategic approach. Leveraging various funding sources, building a compelling case, and networking effectively are critical. By positioning your project as a solution to both agricultural and global challenges, you can attract the investment needed to drive innovation and sustainability in the farming industry.

Investment Opportunities in AgriTech

In an age where artificial intelligence (AI) is transforming industries across the globe, AgriTech stands out as a burgeoning field ripe for investment. The integration of AI in agriculture not only introduces new technologies but also heralds significant improvements in efficiency, productivity, and sustainability. Investors who recognize and capitalize on these opportunities stand to gain substantially, both financially and in terms of contributing to the next green revolution.

The advantages of investing in AgriTech are multifaceted. One of the primary drivers is the exponential growth in global population and the resultant demand for food. By 2050, the world population is expected to reach nearly 10 billion, requiring a 70% increase in food production. Traditional farming methods simply can't keep pace with this demand. Enter AI-driven technologies, which offer innovative solutions to optimize every aspect of agricultural production, from soil analysis to crop management.

Startups and established tech companies are spearheading these innovations, developing tools that harness the power of machine learning, robotics, and data analytics. Investors have a unique opportunity to back these pioneers at various stages of their growth. Early-stage venture capital allows for significant involvement and potentially high returns, albeit with higher risk. Later-stage investments can focus on companies with proven technologies and established market presence, offering more stability and incremental growth.

Precision agriculture is one of the most promising areas for investment within AgriTech. This approach utilizes AI to gather and analyze data from various sources, providing actionable insights that can enhance crop yields and reduce resource wastage. Companies focusing on sensor technology, drones, and IoT (Internet of Things) devices are particularly attractive, as they provide the hardware backbone for data collection and real-time monitoring.

Another critical area is AI-driven crop monitoring and management systems. These platforms leverage satellite imagery, drones, and advanced algorithms to detect diseases, pests, and other crop health issues at an early stage. Investing in this segment not only addresses a crucial need in agriculture but also aligns with the growing trend of sustainable farming practices.

Moreover, soil analysis technologies are rapidly evolving with the help of AI. Traditionally, soil testing has been a labor-intensive and time-consuming process. However, companies are now developing AI models that can predict soil health and suggest amendments in real-time. By investing in these technologies, venture capitalists can contribute to a more efficient and productive agricultural sector, ultimately leading to better yields and reduced environmental impact.

Water management is another compelling sector for AgriTech investment. Given the increasing scarcity of water resources worldwide, AI solutions that optimize irrigation systems and monitor soil moisture levels are gaining traction. These innovations not only save water but also enhance crop quality and yield. Investing in companies that develop AI-powered irrigation systems can be particularly rewarding as they address one of the most pressing challenges facing global agriculture today.

The market for AI-powered autonomous machinery in agriculture is also expanding. From self-driving tractors to robotic harvesters, these machines are designed to perform labor-intensive tasks with high precision and efficiency. Investing in companies that are at the forefront of developing these technologies can yield significant returns, especially as labor shortages continue to affect the agricultural sector.

Predictive maintenance is another area where AI is making waves. By predicting equipment failures before they occur, AI systems can reduce downtime and maintenance costs for farmers. Investors can focus on startups that provide AI-based solutions for machinery upkeep, which enhances operational efficiency and maximizes output.

Furthermore, AI's role in supply chain optimization cannot be overlooked. Companies that use AI to streamline logistics, improve storage conditions, and manage distribution more effectively are transformative. Investment in this realm supports a holistic approach to agriculture that spans the entire value chain, fostering resilience and responsiveness in the market.

Market analysis and decision-making platforms are also critical investment opportunities. These platforms aggregate data from various sources, providing farmers with insights into market trends, price fluctuations, and consumer preferences. AI-driven market analysis tools empower farmers to make informed decisions, enhancing their profitability and reducing risks.

Smart greenhouses that use AI to control environmental conditions represent another lucrative investment opportunity. By optimizing temperature, humidity, and light levels, these greenhouses enable year-round production of high-quality crops. Investing in AI-driven greenhouse technologies can contribute to a more reliable and consistent food supply.

Beyond the practical aspects, the integration of AI in livestock management is also gaining momentum. From tracking animal health to automating feeding and milking processes, AI technologies are revolutionizing livestock farming. Investors who focus on this niche market have the potential to significantly enhance animal welfare and farm productivity.

On a broader scale, AI's influence on promoting sustainability and reducing the environmental impact of agriculture is becoming increasingly important. Investors with a focus on Environmental, Social, and Governance (ESG) criteria will find numerous opportunities in AgriTech. AI technologies that enhance biodiversity, optimize resource use, and minimize carbon footprints are essential for sustainable development.

Despite the promising prospects, investors must also navigate certain economic challenges and considerations. The initial cost of implementing AI technologies can be substantial, and there's often a steep learning curve for farmers. However, the long-term benefits, including cost reduction and efficiency gains, outweigh these initial hurdles. Therefore, patient capital with a long-term vision is essential.

Funding avenues for AgriTech innovations are diverse. Government grants, private venture capital, and corporate investment are all viable options. Collaborations between tech firms, agricultural companies, and research institutions can also foster groundbreaking advancements, creating a robust ecosystem for AgriTech innovation.

In conclusion, the investment opportunities in AgriTech are vast and varied. From precision agriculture and crop monitoring to water management and autonomous machinery, AI is set to revolutionize the farming landscape. Investors who recognize the potential of these technologies not only stand to gain financially but also play a crucial role in shaping the future of sustainable agriculture. The fusion of AI and agriculture is more than just a trend; it's a transformative movement with the potential to address some of the most pressing challenges faced by humanity today. The time to invest is now, as the seeds of the future are already being sown.

Chapter 24:
AI and Climate Change

In an era where climate change poses a formidable challenge to agriculture, AI emerges as a powerful ally in mitigating its impacts and adapting to new environmental realities. By leveraging advanced machine learning algorithms, AI can analyze complex climate patterns and predict extreme weather events, enabling farmers to prepare and protect their crops. AI-powered tools can optimize irrigation systems to conserve water in drought-prone areas and monitor soil health to predict and mitigate the effects of soil degradation. The ability of AI to provide real-time insights and adaptive strategies ensures that farming practices not only survive but thrive in the face of climate change. This seamless integration of AI within agricultural frameworks exemplifies a methodological shift towards more sustainable and resilient farming, setting the stage for a future where technology and nature coexist harmoniously in the quest to secure global food production.

Mitigating Climate Impact through AI

Climate change presents a profound challenge to agriculture, threatening food security and farmer livelihoods across the globe. The erratic weather patterns, increased frequency of extreme weather events, and shifting climatic zones are introducing new variables that traditional farming practices are often ill-equipped to handle. Here, AI steps in as a potent ally, harnessing the power of data and predictive analytics to devise strategies that mitigate climate impact on agriculture.

AI-driven weather forecasting systems are among the most immediate and practical applications helping farmers adapt to climate change. By analyzing vast amounts of historical weather data and current atmospheric conditions, these systems can provide more accurate and localized weather forecasts. These forecasts are crucial for planning planting and harvesting activities, thereby reducing crop losses due to unexpected weather events. It's not just about knowing when it will rain; it's about predicting rain's timing, intensity, and impact on specific plots of land.

Another critical area where AI is making strides is in the optimization of resource use—specifically water. Efficient irrigation systems powered by AI use real-time data from sensors placed in fields to monitor soil moisture levels and weather conditions. These systems can then predict the water needs of various crops and adjust irrigation schedules accordingly. By ensuring that water is used judiciously, AI helps to reduce wastage and maintain the health of water-dependent ecosystems.

Additionally, AI technologies are instrumental in enhancing carbon sequestration practices. By analyzing soil composition and biomass growth patterns, AI systems can recommend agricultural practices that increase the amount of carbon stored in the soil. Techniques such as cover cropping, no-till farming, and crop rotation can be optimized through AI insights, boosting their effectiveness in capturing atmospheric carbon and turning it into organic matter.

AI also plays a substantial role in managing agricultural biodiversity, which is a key factor in building resilience against climate change. Diverse ecosystems tend to be more resilient to pests, diseases, and extreme weather conditions. Through machine learning algorithms, AI can analyze the interaction between different species within an ecosystem, recommending the best crop varieties and planting strategies to promote biodiversity. For example, intercropping—which involves growing two or more crops in proximity—can be optimized using AI to ensure that each plant's growth complements the others, reducing vulnerability to pests and enhancing soil health.

In pest and disease management, AI provides solutions that are both proactive and reactive. AI models can predict pest outbreaks by analyzing climate data, crop conditions, and historical pest behavior. This foresight allows farmers to take preventive measures before an infestation becomes severe, utilizing eco-friendly pest control methods rather than relying heavily on chemical pesticides. Moreover, AI-enabled drones and ground-based robots can identify and treat infested plants with pinpoint accuracy, reducing the overall chemical load on the environment.

The advancements in AI extend to livestock management as well, where they contribute to reducing greenhouse gas emissions from animal agriculture. Machine learning models can optimize feeding regimens to enhance digestion efficiency in livestock, thereby reducing methane emissions. Additionally, AI-driven systems can monitor the health and productivity of animals in real-time, ensuring that they are raised in conditions that minimize stress and maximize health, further contributing to lower emission levels.

AI is also revolutionizing the development of climate-resilient crop varieties. Traditional plant breeding methods, which involve a lot of trial and error, are time-consuming and labor-intensive. AI accelerates this process by analyzing genetic information, environmental conditions, and crop performance data to identify traits associated with climate resilience. These insights enable scientists to develop new crop varieties that can withstand extreme conditions like drought, heat, and floods. It's an ongoing quest for better crops, faster, and AI is at its forefront.

Moreover, supply chain optimization through AI significantly reduces the environmental footprint of agricultural products. By using advanced analytics to enhance logistics, AI helps reduce the distance food travels from farm to table, cutting down on greenhouse gas emissions associated with transportation. Efficient storage and distribution systems enabled by AI also reduce food waste, a critical factor given that food waste contributes significantly to global carbon emissions.

AI applications don't just stop at mitigating adverse impacts; they also offer tools for monitoring and improving the overall sustainability of farming practices. For instance, blockchain technology integrated with AI can provide transparency regarding the sustainability practices at different stages of the supply chain. This ensures that consumers are informed about the environmental impact of their food choices, encouraging more sustainable consumption patterns.

Given the inevitable changes that climate change will bring to agricultural landscapes, adapting to these changes is as crucial as mitigating them. AI helps farmers to adapt by offering predictive analytics and scenario planning. By simulating various climate scenarios, AI can help farmers make long-term decisions about which crops to plant, when to plant them, and how to manage their fields to weather the storms of climate change. It's about anticipating the future and preparing for it, ensuring that agriculture remains viable even in the face of unprecedented challenges.

One of the more inspiring aspects of using AI to combat climate change in agriculture is the potential for these technologies to democratize access to advanced farming techniques. While large-scale farms may have the resources to experiment with new technologies, small-scale farmers often don't. AI-powered mobile apps and platforms provide smallholder farmers with access to the same sophisticated tools used by larger farms, helping them make data-driven decisions that can improve their yield and sustainability. These tools often come in the form of user-friendly apps, making cutting-edge technology accessible to those who need it the most.

The collaborative nature of AI research and development also means that different stakeholders—governments, tech companies, academic institutions, and non-profits—are working together to create solutions tailored to diverse agricultural contexts. Governments can set favorable policies and provide funding for AI research in agriculture, while tech companies can offer the necessary computational power and expertise. Academic institutions can contribute through research and development, and non-profits can work on making these technologies accessible to farmers around the world. This multi-stakeholder approach ensures that AI solutions are both innovative and inclusive.

It's vital to note that while AI offers transformative solutions, it needs to be deployed thoughtfully to avoid potential pitfalls. Ethical considerations surrounding data privacy, algorithms' transparency, and the potential displacement of labor are critical factors that must be addressed. However, when implemented responsibly, AI holds the promise to not only mitigate the impacts of climate change on agriculture but to transform how we think about farming in a climate-challenged world.

In summary, AI brings a suite of tools and techniques that can fundamentally change the way we approach agriculture in the era of climate change. From optimizing resource use and developing resilient crops to enhancing biodiversity and reducing carbon emissions, the possibilities are vast and transformative. While challenges remain, the path ahead is promising, driven by the convergence of innovation and sustainability. In embracing AI, we are not just combating climate change; we are paving the way for a new era of sustainable and resilient agriculture.

Adapting to Changing Environmental Conditions

As the global climate continues to shift unpredictably, farmers find themselves facing a slew of challenges that can make traditional agricultural practices seem archaic and inefficient. The irregular patterns of rainfall, the increased frequency of extreme weather events, and fluctuating temperatures all contribute to an environment where consistency is a rare commodity. It's within this context that AI's potential truly shines, offering innovative solutions to adapt and thrive under these changing conditions.

One of the most straightforward applications of AI in this sphere is in weather prediction. Traditional methods of forecasting weather often fall short in their capacity to provide the level of precision required for making informed agricultural decisions. AI systems, with their ability to analyze vast datasets and recognize subtle patterns, can offer more accurate and timely forecasts. Farmers equipped with these insights can make more informed decisions about planting, harvesting, and even applying fertilizers or pesticides. This predictive ability isn't just about convenience; it's a matter of survival and optimization in an increasingly uncertain world.

Moreover, AI isn't limited to merely forecasting weather conditions. It dives deeper, analyzing climatic trends over time to offer long-term projections. This means farmers can prepare in advance for potential droughts, floods, or unexpected frosts. The ability to look several seasons ahead and adjust farming practices accordingly is revolutionary. For instance, farmers can adopt different crop varieties that are better suited to anticipated conditions, or they might decide to implement new water management strategies ahead of a predicted drought.

In the face of erratic climate patterns, soil health becomes an even more crucial factor in successful farming. AI-driven platforms can help farmers constantly monitor soil conditions, detecting changes that might be detrimental to crop health. For example, using AI to analyze soil moisture levels can lead to smarter irrigation practices. By only using water when and where it's needed, farmers can conserve a precious resource, reduce costs, and maintain optimal soil conditions.

Crop resilience is another vital factor when adapting to climate change. AI can assist in breeding programs to develop new crop varieties that are more resilient to extreme weather conditions. Machine learning algorithms can quickly analyze genetic data to identify traits associated with higher tolerance to drought, heat, or disease. These resilient crops are better suited to withstand the kinds of environmental stressors that are becoming more common.

Beyond soil and crop management, AI has a role in pest and disease management, which are becoming more unpredictable due to climate change. Warmer temperatures and changing precipitation patterns can lead to the emergence of new pests and diseases or alter the lifecycle of existing ones. AI systems that analyze environmental data alongside pest and disease data can predict outbreaks before they occur. Farmers can receive timely alerts and take preemptive action, whether that's deploying biological controls or applying targeted pesticides. This proactive approach is both more effective and more sustainable.

Water management presents another significant challenge in the context of climate change. AI-driven irrigation systems can optimize water usage by delivering just the right amount of water to the crops, thereby reducing waste and enhancing crop yield. These systems use sensors and IoT devices to gather data on soil moisture, weather conditions, and crop requirements. AI algorithms then process this data in real-time, allowing for precise and dynamic irrigation scheduling, which is especially crucial during periods of drought or water scarcity.

Farmers are also leveraging AI to manage the health of their livestock under changing environmental conditions. AI can monitor animal health using wearable devices that track movement patterns, body temperature, and other vital signs. By analyzing this data, AI can detect early signs of stress or illness, which are often exacerbated by fluctuating environmental conditions. Early intervention leads to healthier animals and more efficient farm operations.

The logistics of farming also need to adapt to changing environmental conditions. AI plays a vital role in optimizing supply chain operations through predictive analytics. By forecasting demand and managing inventory more effectively, AI can ensure that produce reaches markets without unnecessary delays or spoilage, despite environmental disruptions. This level of efficiency is essential for minimizing waste and ensuring food security.

It's worth noting that the application of AI in adapting to climate change is not limited to large-scale farming. Smallholder farmers, who are often the most vulnerable to climate change, can also benefit from AI technologies. Mobile-based AI applications can provide small-scale farmers with accessible tools for weather forecasting, soil analysis, and even market pricing. These tools empower them to make informed decisions, improving their resilience to climate variability.

Integrating AI into agricultural practices requires a mindset change as well as substantial investments in technology and infrastructure. Governments, NGOs, and private sector stakeholders must collaborate to support farmers through this transition. Subsidies for AI technologies, training programs, and development of localized AI-driven solutions are essential in making this transition a reality for farmers worldwide.

The road ahead will certainly have its challenges, but the potential rewards are immense. By harnessing the power of AI, farmers can transform adversity into opportunity, turning the unpredictable nature of climate change into a manageable aspect of their operations. This is not merely about adapting to survive; it's about thriving in a new agricultural paradigm where data and innovation lead the way.

In conclusion, AI's role in helping agriculture adapt to changing environmental conditions is multi-faceted and far-reaching. From accurate weather predictions and efficient water management to resilience-building and advanced pest control, the applications are as diverse as they are impactful. This synergy of technology and farming holds promise not just for surviving but flourishing in the face of climate adversity. As we look to the future, it becomes clear that AI will be an indispensable tool in the quest for sustainable and resilient agricultural practices.

Chapter 25:
Community Engagement
and Social Impact

Embracing artificial intelligence in agriculture isn't just about technological advancement; it's about fostering a sense of community and generating widespread social benefits. By integrating AI solutions tailored to the unique needs of local farmers, communities can unlock new opportunities for collaboration and economic resilience. Farmers' collectives can leverage predictive analytics to plan crop rotations and share resources more effectively, reducing wastage and boosting productivity. AI-driven platforms can facilitate knowledge transfer between experienced and novice farmers, ensuring sustainable practices are passed down and adapted to local conditions. Additionally, community engagement in the development and deployment of these technologies ensures that AI tools are not only accessible but also culturally relevant, promoting inclusive growth. Ultimately, the social impact of AI in agriculture extends beyond individual farms, fostering a culture of innovation and collective progress that can transform entire rural economies for the better.

Engaging Local Communities

Engaging local communities in the advancements of AI-powered agriculture isn't just about implementing cutting-edge technologies; it's about ensuring that the benefits of these innovations reach the very grassroots. For AI to genuinely revolutionize agriculture, it must resonate with the farmers, laborers, and stakeholders who are the backbone of this industry. By fostering engagement and participation at the local level, we essentially create a bridge between futuristic technology and traditional agricultural practices, ensuring that the transformation is both inclusive and substantial.

One of the most effective ways to engage local communities is by involving them in the decision-making process from the get-go. Farmers, often steeped in generations of agricultural knowledge, can provide invaluable insights that can complement and enhance AI solutions. Hosting community meetings, workshops, and discussions can facilitate an exchange of ideas, allowing tech developers to tailor their solutions to the specific needs and conditions of the local environment. For instance, understanding the particular pests, crop cycles, soil types, and weather patterns of an area can significantly refine the efficacy of AI models.

Equally important is the aspect of education and training. AI in agriculture can seem daunting and out of reach for many farmers, especially those in rural areas with limited access to technology. Comprehensive training programs are essential in demystifying AI and making it more accessible. These programs should not only focus on how to use AI tools but also on understanding the underlying principles of these technologies. Visual aids, hands-on demonstrations, and local language resources can be highly effective in this regard.

Collaboration with local educational institutions can enhance these efforts. Schools, colleges, and agricultural training centers can integrate AI into their curriculums, fostering a new generation of tech-savvy farmers who are comfortable with both traditional methods and advanced technologies. This symbiotic relationship can also extend to involve local tech entrepreneurs who can develop customized AI solutions addressing local issues, thus fostering a community-driven innovation ecosystem.

Moreover, building infrastructure that supports the use of AI in agriculture can significantly boost community engagement. Reliable internet connectivity, access to smart devices, and local support centers are critical. These facilities ensure that farmers can leverage AI technologies efficiently. Government and non-profit organizations can play a crucial role here by funding and setting up these infrastructures in underserved areas.

Trust plays a pivotal role in the adoption of new technologies. Farmers need to trust that AI solutions will indeed improve their yields, reduce their costs, and ultimately benefit their livelihoods. Building this trust requires transparency and consistent demonstration of positive outcomes. Pilot projects and small-scale implementations can serve as proof of concept, showcasing tangible benefits and creating a ripple effect that can encourage wider adoption.

Financial incentives and support mechanisms can also drive engagement. Subsidies, grants, and low-interest loans for AI-related agricultural tools can reduce the economic barriers for farmers. Microfinancing institutions and agri-tech startups can collaborate to provide financial packages tailored to small and marginal farmers. This financial empowerment can significantly boost the adoption rates of AI technologies.

Gender inclusivity is another crucial aspect of community engagement. Women form a substantial part of the agricultural workforce in many parts of the world, yet they often have limited access to technology and resources. Specific initiatives aimed at training and supporting women farmers in the use of AI can bridge this gap, fostering a more inclusive and equitable agricultural transformation.

One can't overlook the social benefits that arise from engaging local communities with AI in agriculture. Beyond the economic gains, there's a potential for improved social cohesion and community spirit. Shared experiences with new technologies can forge stronger community bonds. Farmers can form cooperative groups, sharing resources and knowledge, which can also lead to collective bargaining and better market access.

Success stories and testimonials from local farmers who have benefited from AI implementations can serve as powerful motivators. Case studies showcasing increased yields, reduced labor, and better market prices thanks to AI adoption can spark interest and willingness among other community members to embrace these technologies. Such stories humanize the technology, making it relatable and attainable.

It's also essential to listen to feedback and continuously refine AI tools to better meet the needs of the farming community. An iterative approach, where technology evolves based on real-world usage and feedback, can significantly enhance relevance and adoption. This approach helps in creating AI solutions that are not only high-tech but also high-touch, reflecting the realities and challenges of everyday farming.

In conclusion, the engagement of local communities in the AI-driven agricultural revolution is vital. It ensures that the technology isn't just an external force but an integrated part of the agricultural ecosystem. By involving farmers in the decision-making process, providing education and training, building supportive infrastructure, fostering trust, offering financial incentives, and ensuring inclusivity, we can create a transformation that is not just technologically advanced but also socially rooted and widely beneficial. The journey of AI in agriculture is a shared one, and it's through community engagement that this journey will find its most successful and sustainable path.

Social Benefits of AI in Agriculture

Artificial Intelligence (AI) is revolutionizing many aspects of our lives, and its influence is profoundly felt in agriculture. But beyond driving efficiencies and boosting yields, the integration of AI into agriculture brings significant social benefits that contribute to community well-being and societal progress. One of the most immediate impacts is on local job creation and skill development. As AI technologies become more prevalent, they necessitate the emergence of new roles - from AI training experts to engineers specialized in agricultural applications. This shift not only creates job opportunities but also fosters an environment where continuous learning and innovation thrive.

AI-enabled precision agriculture allows farmers to optimize the use of resources such as water, fertilizers, and pesticides, which leads to more sustainable farming practices. This has tremendous social implications as it promotes environmental stewardship and ensures that future generations inherit a healthier planet. By reducing chemical usage, AI directly contributes to improved public health in farming communities. Lesser exposure to harmful substances means fewer health problems among farmers and their families, creating a healthier and more productive populace.

Further, the implementation of AI in agriculture can play a crucial role in regional food security. Advanced AI systems are capable of monitoring and predicting crop health, thereby reducing the risk of crop failures. This stability in food production is vital for communities, especially in areas prone to climatic variabilities. By ensuring consistent and reliable food supplies, AI helps mitigate hunger and malnutrition, which are critical social issues in many parts of the world. Those who were previously dependent on unpredictable manual farming techniques can now count on more stable livelihoods.

Another transformative social benefit of AI in agriculture is the empowerment of smallholder farmers. These farmers often face numerous challenges, including limited access to technology, markets, and financial services. AI-driven platforms can bridge these gaps by providing small farmers with real-time data, analytic insights, and access to broader markets. For instance, predictive analytics can inform farmers about the best times to plant and harvest, while market forecasting tools can help them decide when to sell their produce for maximum profit. This democratization of technology ensures that even the smallest of farmers can compete on more equal footing with larger agricultural enterprises.

Moreover, AI in agriculture encourages community-based solutions. Farmers can collectively use AI tools and share data to maximize benefits. This sort of collaboration fosters a stronger sense of community and mutual support, where collective intelligence is built over individual experiences. By working together and leveraging AI, farming communities can solve local agricultural problems more efficiently and innovatively, reinforcing the social fabric that binds them together.

Education and awareness campaigns have also seen a boost from AI technologies. As farmers gain access to AI tools, they simultaneously receive education about sustainable farming practices and technological applications. This education goes beyond the farm; it spreads to the wider community, encouraging everyone to adopt more sustainable and health-conscious habits. The ripple effect of this knowledge-sharing can lead to broader societal changes that extend well beyond agricultural practices.

The social benefits of AI in agriculture also extend to inclusivity. Traditional farming can be physically demanding, often excluding those who are elderly, disabled, or unable to perform strenuous tasks. AI technologies, such as autonomous machinery and robots, lower these barriers, enabling a more diverse population to participate in agricultural activities. This inclusiveness not only enriches the farming community but also ensures a broader spectrum of perspectives and ideas, fostering innovation.

Additionally, AI has the potential to strengthen rural economies. Increased agricultural productivity and efficiency can lead to higher incomes for farmers, which in turn stimulates local economies. With more disposable income, families can invest in better education, healthcare, and infrastructure improvements. The resulting economic upliftment can lead to a higher overall quality of life for rural populations, creating a more vibrant and dynamic community.

Social impact through AI in agriculture is not confined to economic gains and job creation; it also brings about significant cultural changes. The advent of AI fosters a culture of innovation and forward-thinking within agricultural communities. Young people, who might have previously considered farming a less attractive career, now see it as a field buzzing with advanced technology and opportunities for growth. This shift can help mitigate the problem of urban migration, encouraging younger generations to stay and contribute to rural communities.

Moreover, AI's role in enhancing transparency and traceability in agriculture can lead to more ethical farming practices. With better tracking of the supply chain, consumers gain confidence in the origin and quality of their food. This transparency can prompt farmers to engage in more ethical practices, knowing that their methods and processes are being monitored and valued by the end consumer. Such accountability can foster a more ethically conscious farming community and promote fairness and integrity in agricultural production.

Significant also is the psychological impact of AI in agriculture on farming communities. The introduction of AI tools can reduce the stress and uncertainty that come with traditional farming methods. With predictive analytics and automated systems, farmers can make more informed decisions, reducing the guesswork and anxiety that often accompany agricultural activities. This mental ease translates to a happier, more motivated, and resilient farming community.

When looking towards the global implications, AI in agriculture offers a vital platform for international collaborations. Knowledge-sharing across borders can lead to innovative solutions to common agricultural challenges, fostering a sense of global community. Farmers in developing countries can learn from advancements in more developed regions, ensuring that insights and technologies are shared widely and inclusively. This global solidarity can help improve agricultural practices on a large scale, driving collective progress and societal development.

The final critical element of the social benefits of AI in agriculture is how it can reshape the narrative around farming. As AI technology makes farming more attractive, younger individuals may be more inclined to pursue careers in agriculture. The infusion of fresh perspectives can result in new, innovative solutions that continue to advance the field. By aligning farming with cutting-edge technology, AI helps depict agriculture as a forward-thinking and vital industry, essential for sustainable development and social well-being.

In conclusion, AI in agriculture offers a myriad of social benefits that stretch far beyond productivity and yield. By fostering job creation, enhancing sustainability, ensuring food security, empowering smallholder farmers, promoting inclusivity, boosting rural economies, and catalyzing cultural shifts, AI stands as a pivotal force for social good. Its transformative potential lies in its capacity to bring about comprehensive societal benefits, making agriculture a cornerstone of sustainable and inclusive progress.

Conclusion

As we stand on the precipice of a new era in agriculture, driven by artificial intelligence, it becomes apparent that the journey has just begun. AI's transformative potential reaches far beyond simple automation and efficiency. It's paving the way for smarter decisions, more sustainable practices, and greater yields that can feed a growing global population. Through the chapters of this book, we've explored the multifaceted roles and applications of AI in farming, each innovation holding the promise of revolutionizing how we grow, monitor, and manage our crops and livestock.

The dawn of AI in agriculture is not just a technological milestone; it's a call to reimagine traditional agricultural practices. As farmers and technologists unite, the synergy between human intuition and machine intelligence is creating outcomes that were once dreams. Fields monitored in real-time, soil health managed with precision, and water resources utilized optimally showcase just a fraction of AI's capabilities. These advancements signify a paradigm shift in how we address one of humanity's oldest challenges: producing enough food while safeguarding the environment.

Moreover, the role of AI extends into the heart of sustainability. In the face of climate change and depleting natural resources, AI provides tools to reduce carbon footprints, promote biodiversity, and ensure conservation efforts are data-driven and effective. Precision agriculture's capacity to utilize resources judiciously has the dual effect of improving yields and preserving the natural ecosystem. The narrative of AI and sustainability isn't just about survival; it's about thriving in harmony with our planet.

The economic implications are equally profound. AI's ability to reduce costs, increase efficiencies, and open new avenues for revenue can level the playing field, especially for smallholder farmers. By democratizing access to advanced tools and knowledge, AI is making it possible for even the smallest farms to compete globally. The economic case for AI in agriculture is strong, promising to enhance both profitability and sustainability in tandem.

However, it's essential to acknowledge that the road to widespread AI adoption is not without obstacles. From technological limitations to skepticism and the need for robust policies, the challenges are varied. Yet, these hurdles present opportunities for growth and improvement. Addressing these issues head-on will be crucial for the seamless integration of AI in agriculture. As with any revolutionary technology, continuous learning, adaptation, and collaboration will be paramount to overcoming these barriers.

One of the most significant leaps will be educating the next generation of farmers, technologists, and policymakers. Comprehensive training programs, workshops, and curricula focused on AI in agriculture will ensure that the workforce is well-equipped to harness the potential of these technologies. As AI continues to evolve, so too must our educational frameworks, enabling a seamless transition from traditional to tech-savvy farming practices.

In conclusion, AI's role in agriculture represents a confluence of innovation, sustainability, and economic viability. The insights gleaned from various case studies across the globe reinforce the notion that AI's impact transcends borders and cultural contexts. The future trends in AI agriculture look promising, with emerging technologies and collaborations poised to further the cause of a sustainable and productive agricultural sector.

This book has illustrated that AI in agriculture is more than just a technological advance; it's a testament to human ingenuity and the relentless pursuit of progress. As we collectively navigate the complexities and embrace the opportunities, we are setting the stage for a future where technology and nature coexist in a symbiotic relationship. The next green revolution is on the horizon, and AI is at its helm. The journey ahead is promising, and the possibilities are endless. Let's embark on this transformative path towards a more sustainable, efficient, and prosperous agricultural future.

Appendix A:
Appendix

The Appendix serves as a comprehensive resource providing supplementary information that complements the main content of this book. Here, readers will find detailed references, data sets, formulas, and additional resources that are integral for a deeper understanding of the transformational impact of artificial intelligence in agriculture. While the chapters introduce and elucidate concepts, strategies, and case studies, this appendix offers tools and resources for applying this knowledge practically.

Data Sets and References

In undertaking the journey of integrating AI into agricultural practices, access to reliable data sets is crucial. This section lists various publicly available data sources, including crop yield databases, weather patterns, soil health records, and market trends. These data sets can be invaluable for building and training AI models tailored to specific needs.

- National Agricultural Statistics Service (NASS)

- FAO Statistical Database (FAOSTAT)

- Open Agriculture Data Alliance (OADA)

- Local and regional agricultural departments and research centers

Formulas and Equations

This section contains key formulas and equations used in precision agriculture and AI models. Whether it's calculating the optimal irrigation levels or analyzing soil nutrient content, these formulas are fundamental tools for farmers and developers alike.

- Soil Moisture Balance Equation

- Yield Prediction Algorithms

- Evapotranspiration Formula

- Pest Population Growth Models

Software Tools and Applications

Software and hardware tools form the backbone of any AI-driven agricultural solution. This section outlines various software frameworks, AI platforms, and open-source tools that facilitate the development and deployment of agricultural AI applications.

- TensorFlow and PyTorch for machine learning model building

- OpenCV for image processing applications in crop monitoring

- GIS software for spatial data analysis

- Farm management software like AgriTech and John Deere Operations Center

Additional Reading and Resources

To foster a deeper understanding and inspire further exploration, this section lists additional reading materials and resources. These include scholarly articles, research papers, journals, and books by experts in the disciplines of artificial intelligence, agronomy, and sustainable farming.

- "Artificial Intelligence in Agriculture" by Ajith Abraham
- "Precision Agriculture Technology and Economic Perspectives" by Steve Smyth
- Journals like "Computers and Electronics in Agriculture"
- Educational websites and online courses from institutions like MIT and Stanford

Contact Information

For readers interested in connecting with industry experts, research institutions, or collaborative projects, contact information is provided. This includes email addresses, websites, and social media handles of relevant organizations and thought leaders in the field of AI and agriculture.

- National Institute of Food and Agriculture (NIFA)
- International Society of Precision Agriculture (ISPA)
- Farmers Business Network (FBN)
- Social media groups and forums on LinkedIn and Reddit

Glossary of Terms

This glossary provides definitions and explanations of key terms used throughout the book. These terms are essential for understanding how artificial intelligence (AI) is transforming agriculture and promoting sustainable farming practices.

Artificial Intelligence (AI)

A branch of computer science focused on creating systems capable of performing tasks that typically require human intelligence, such as visual perception, speech recognition, decision-making, and language translation.

Algorithm

A step-by-step procedure or formula for solving a problem. In the context of AI, algorithms are sets of rules or instructions that help machines learn and make decisions.

Autonomous Machinery

Farming equipment, such as tractors and harvesters, that operate with minimal human intervention using AI for navigation, task execution, and real-time decision-making.

Big Data

Large volumes of data that can be analyzed to reveal patterns, trends, and associations, especially relating to human behavior and interactions. In agriculture, big data includes information from sensors, satellite imagery, weather data, and more.

Blockchain

A decentralized digital ledger that records transactions across many computers in a way that makes the records secure and transparent. It is used in agriculture to enhance traceability and transparency in the supply chain.

Deep Learning

A subset of machine learning that uses neural networks with many layers (hence "deep") to analyze various factors of data. It is particularly effective for tasks like image and speech recognition.

Ecosystem

A biological community of interacting organisms and their physical environment. In agriculture, maintaining a balanced ecosystem is vital for sustainable farming.

Internet of Things (IoT)

A network of interconnected devices that communicate and exchange data with each other. In agriculture, IoT devices include sensors and drones used for monitoring crops and soil conditions.

Machine Learning (ML)

A subset of AI that enables computers to learn from and make decisions based on data without being explicitly programmed. Machine learning algorithms improve over time as they are exposed to more data.

Neural Network

A computing system inspired by the human brain's network of neurons. Neural networks are used in AI to recognize patterns and make decisions based on data inputs.

Precision Agriculture

A farming management concept that uses technology to observe, measure, and respond to variability in crops. It includes the use of GPS, IoT, and AI to optimize field-level management regarding crop farming.

Remote Sensing

The acquisition of information about an object or area from a distance, typically from aircraft or satellites. In agriculture, remote sensing monitors crop health, soil conditions, and weather patterns.

Robotics

The branch of technology that deals with the design, construction, operation, and application of robots. In agriculture, robots are used for tasks like planting, weeding, and harvesting crops.

Sustainability

Practices that meet current agricultural needs without compromising the ability of future generations to meet theirs. It includes conserving resources, reducing pollution, and promoting biodiversity.

UAV (Unmanned Aerial Vehicle)

Commonly known as drones, UAVs are used in agriculture for aerial imaging, crop monitoring, and spraying fertilizers or pesticides.

Variable Rate Technology (VRT)

Technology that allows the variable application of inputs like fertilizers, seeds, and water. VRT systems use GPS and other data to apply the right amount of input in the right place at the right time.

Yield Mapping

The process of collecting georeferenced data on crop yield and moisture content during harvest. This data helps farmers analyze and understand the variability within their fields, leading to better crop management.

By understanding these terms, you'll be better equipped to grasp the concepts and technologies discussed in this book. This knowledge will empower you to see the vast potential of AI in revolutionizing agriculture, making it more efficient and sustainable.

List of Resources and Further Reading

For those interested in diving deeper into the transformative role of AI in agriculture, an extensive range of resources is available. These materials cover theoretical insights, real-world applications, and future possibilities, helping both newcomers and experienced practitioners expand their understanding of this dynamic field.

First and foremost, a wealth of scholarly articles and journals provide in-depth analyses of various AI technologies in agriculture. Academic databases such as IEEE Xplore, ScienceDirect, and Springer offer access to papers that investigate everything from AI-driven soil health analysis to the predictive capabilities of machine learning algorithms in crop management. Researchers can find detailed methodologies, experimental results, and discussions on the implications of these technologies, making these sources invaluable for academic and field research.

Books are another excellent resource, offering comprehensive overviews and in-depth discussions. Titles like "Artificial Intelligence: A Modern Approach" by Stuart Russell and Peter Norvig provide a thorough grounding in AI principles, while "AI in Agriculture" by Raj Biredj and others focuses on specific applications in the agricultural sector. These books often include case studies that illustrate successful implementations and highlight the challenges faced along the way.

Online platforms and courses can help those looking to gain practical skills in applying AI to agriculture. Websites like Coursera, Udemy, and edX offer courses on machine learning, data science, and IoT applications in farming. These courses often feature interactive exercises and projects, allowing learners to apply theoretical knowledge in simulated or real-world scenarios. Such hands-on experience can be crucial for anyone looking to leverage AI technologies on their farms or within agricultural enterprises.

Industry reports and white papers from organizations such as the Food and Agriculture Organization (FAO) of the United Nations, McKinsey & Company, and Boston Consulting Group offer insights into global trends and market dynamics. These documents often explore the economic viability of AI technologies, discuss policy implications, and provide recommendations for stakeholders. Businesses, policymakers, and academic institutions can benefit from these comprehensive analyses to inform their strategies and decision-making processes.

Several reputable websites and blogs focus exclusively on innovations in AgriTech and AI. Websites like AgFunder, Agri-Tech East, and FarmBot provide news, articles, and opinion pieces on the latest developments in the field. Keeping up with these sources can help readers stay informed about emerging technologies, track startup activity, and understand investment trends. Additionally, engaging with the comments sections or community forums on these websites can provide opportunities for networking and knowledge exchange.

Conferences and webinars are another great way to stay updated and connect with experts. Events organized by groups such as the International Conference on Smart Agriculture Technologies (ICSAT), and the Precision Agriculture Symposium, often feature keynote speeches, panel discussions, and workshops led by industry leaders and researchers. Attending these events can provide first-hand insights into cutting-edge research, new product launches, and upcoming trends. Many of these events are now available online, making them more accessible than ever.

Government and NGO publications also contribute significantly to the pool of knowledge. Reports and policy briefs by entities like the United States Department of Agriculture (USDA), European Commission, and various other regional agricultural departments can help readers understand the regulatory landscape and funding opportunities available for AI projects in agriculture. These publications often include case studies and pilot project results, providing real-world examples of AI applications in different agricultural contexts.

Furthermore, there are numerous open-source datasets and platforms that allow for practical experimentation and analysis. Websites such as Kaggle, GitHub, and the UCI Machine Learning Repository offer datasets related to crop yields, soil conditions, weather patterns, and pest infestations. These datasets can be used for developing and testing AI models, offering a hands-on approach to learning and innovating. Tutorials and documentation available on these platforms can guide users through the process of data acquisition, model training, and performance evaluation.

Podcasts and video channels also provide valuable information in an easily digestible format. Channels like "The Future of Agriculture" podcast and YouTube series like "Agri-Tech Future" explore various aspects of AI in farming through interviews with experts, success stories, and discussions on future trends. These multimedia resources can be particularly useful for those who prefer to consume information on the go or in a more conversational format.

Lastly, connecting with professional organizations and societies can provide ongoing support and resources. Groups such as the American Society of Agricultural and Biological Engineers (ASABE) and the International Society of Precision Agriculture (ISPA) offer memberships that include access to journals, newsletters, and networking events. Being part of a professional community can offer continued learning opportunities, peer support, and collaboration possibilities.

By leveraging these diverse resources, individuals, farmers, and tech enthusiasts can deepen their knowledge, stay updated on the latest developments, and actively participate in the burgeoning field of AI in agriculture. Whether through academic research, practical courses, industry insights, or community engagement, there's a wealth of information available to inspire and educate those ready to embrace the next green revolution.

Glossary of Terms: Acknowledgments

This book wouldn't have been possible without the contributions and support of numerous individuals and organizations. Our deepest gratitude goes to the farmers and agricultural experts who shared their invaluable experiences and insights. Their real-world knowledge provided the foundation for many of the concepts discussed in this book, bringing a level of depth and practicality that theory alone could never achieve.

We also extend heartfelt thanks to the AI and machine learning researchers whose groundbreaking work led to the development of the technologies explored in these pages. Their relentless pursuit of innovation is shaping the future of agriculture and providing the tools needed to meet the challenges of a rapidly changing world. Specifically, the technical contributions from the AI community have been instrumental in detailing how these technologies can be applied to farming practices, revolutionizing the way we think about agriculture.

A special acknowledgment goes to the academic institutions and research centers that have been at the forefront of integrating AI into agricultural studies. Their contributions to understanding both the theoretical and practical applications of AI are critical to the advancement of this field. The collaborative research efforts between universities and the agricultural industry have provided robust data and case studies, which have enriched the content of this book.

We are equally indebted to the technology companies and startups that are driving innovation in agri-tech. Their investment in research and development has yielded cutting-edge tools and solutions, making AI accessible and practical for everyday farming. These companies' focus on user-centric designs has also ensured that the technologies are not just advanced, but also usable by non-technical farmers.

Government bodies and non-governmental organizations (NGOs) deserve profound appreciation for their role in promoting the adoption of AI in agriculture. Through funding, policy-making, and grassroots initiatives, they have facilitated the integration of innovative technologies in farming communities around the world. Their efforts to bridge the gap between technology and agriculture have been invaluable in bringing about the transformative changes described in this book.

Our deepest gratitude also goes to the editors and technical reviewers who meticulously combed through the drafts, ensuring the accuracy and clarity of the content. Their keen eyes and thorough understanding of both AI and agricultural practices were crucial in refining the concepts presented, guaranteeing that each chapter is both informative and engaging.

Acknowledgments are also due to the contributors who provided real-world case studies, detailing successful implementations of AI in various agricultural settings. These testimonies serve as both proof and inspiration, illustrating a multitude of ways that AI can bring tangible benefits to farming operations. Without their willingness to share experiences, this book wouldn't have the rich, illustrative examples that help bring the text to life.

We also thank the farmers and agricultural cooperatives who allowed us to study their practices and provided detailed feedback on the applicability of AI solutions. Their openness to innovation and practical insights have played a crucial role in shaping the discussions and recommendations in this book.

Moreover, special thanks go to the early adopters and pioneers in the field who faced the initial challenges of integrating AI into farming. Their bold steps and perseverance have paved the way for broader adoption and have laid down the blueprint for others to follow. The stories of their successes and learning experiences serve as a roadmap for the next generation of tech-savvy farmers.

We are also grateful to the peer reviewers who shared their expertise pro bono. Their constructive feedback was invaluable in ensuring the book's rigor and relevance. Through their critical assessment, countless enhancements were made, aligning the content with the current state-of-the-art in both AI and agricultural sciences.

Acknowledgment is due to the grant agencies and funding institutions that supported the initial research efforts leading to significant breakthroughs in AI applications for agriculture. Their financial backing allowed for extensive research and pilot projects that significantly contributed to the information and data presented in this book.

Finally, our deepest gratitude goes to the families and friends who supported the authors and contributors throughout this project. Your encouragement, patience, and unwavering belief in the importance of this work have been the invisible hand guiding it to completion.

In closing, we wish to recognize the countless other unsung heroes whose contributions—no matter how big or small—have played a part in the realization of this book. Together, we are fostering a future where AI and agriculture collaborate to feed the world sustainably and efficiently. It's not just about technology; it's about people working together to create a verdant future.

www.ingramcontent.com/pod-product-compliance
Lightning Source LLC
Chambersburg PA
CBHW051232050326
40689CB00007B/899